HELPING MINDS MEET
Skills for a better life with your dog

Helen Zulch & Daniel Mills

Photography by Peter Baumber

Hubble & Hattie

The Hubble & Hattie imprint was launched in 2009 and is named in memory of two very special Westie sisters owned by Veloce's proprietors.

Since the first book, many more have been added to the list, all with the same underlying objective: to be of real benefit to the species they cover, at the same time promoting compassion, understanding and respect between all animals (including human ones!)

Hubble & Hattie is the home of a range of books that cover all things animal, produced to the same high quality of content and presentation as our motoring books, and offering the same great value for money.

More great Hubble & Hattie books!

Among the Wolves: Memoirs of a wolf handler (Shelbourne)
Animal Grief: How animals mourn (Alderton)
Babies, kids and dogs – creating a safe and harmonious relationship (Fallon & Davenport)
Because this is our home ... the story of a cat's progress (Bowes)
Bonds – Capturing the special relationship that dogs share with their people (Cukuraite & Pais)
Camper vans, ex-pats & Spanish Hounds: from road trip to rescue – the strays of Spain (Coates & Morris)
Canine aggression – how kindness and compassion saved Calgacus (McLennan)
Cat and Dog Health, The Complete Book of (Hansen)
Cat Speak: recognising & understanding behaviour (Rauth-Widmann)
Charlie – The dog who came in from the wild (Tenzin-Dolma)
Clever dog! Life lessons from the world's most successful animal (O'Meara)
Complete Dog Massage Manual, The – Gentle Dog Care (Robertson)
Confessions of a veterinary nurse: paws, claws and puppy dog tails (Ison)
Detector Dog – A Talking Dogs Scentwork Manual (Mackinnon)
Dieting with my dog: one busy life, two full figures ... and unconditional love (Frezon)
Dinner with Rover: delicious, nutritious meals for you and your dog to share (Paton-Ayre)
Dog Cookies: healthy, allergen-free treat recipes for your dog (Schöps)
Dog-friendly gardening: creating a safe haven for you and your dog (Bush)
Dog Games – stimulating play to entertain your dog and you (Blenski)
Dog Relax – relaxed dogs, relaxed owners (Pilguj)
Dog Speak: recognising & understanding behaviour (Blenski)
Dogs just wanna have Fun! Picture this: dogs at play (Murphy)
Dogs on Wheels: travelling with your canine companion (Mort)
Emergency First Aid for dogs: at home and away Revised Edition (Bucksch)
Exercising your puppy: a gentle & natural approach – Gentle Dog Care (Robertson & Pope)

For the love of Scout: promises to a small dog (Ison)
Fun and Games for Cats (Seidl)
Gods, ghosts, and black dogs – the fascinating folklore and mythology of dogs (Coren)
Helping minds meet – skills for a better life with your dog (Zulch & Mills)
Home alone – and happy! Essential life skills for preventing separation anxiety in dogs and puppies (Mallatratt)
Know Your Dog – The guide to a beautiful relationship (Birmelin)
Letting in the dog: opening hearts and minds to a deeper understanding (Blocker)
Life skills for puppies – laying the foundation for a loving, lasting relationship (Zuch & Mills)
Lily: One in a million! A miracle of survival (Hamilton)
Living with an Older Dog – Gentle Dog Care (Alderton & Hall)
Miaow! Cats really are nicer than people! (Moore)
Mike&Scrabble – A guide to training your new Human (Dicks & Scrabble)
Mike&Scrabble Too – Further tips on training your Human (Dicks & Scrabble)
My cat has arthritis – but lives life to the full! (Carrick)
My dog has arthritis – but lives life to the full! (Carrick)
My dog has cruciate ligament injury – but lives life to the full! (Häusler & Friedrich)
My dog has epilepsy – but lives life to the full! (Carrick)
My dog has hip dysplasia – but lives life to the full! (Häusler & Friedrich)
My dog is blind – but lives life to the full! (Horsky)
My dog is deaf – but lives life to the full! (Willms)
My Dog, my Friend: heart-warming tales of canine companionship from celebrities and other extraordinary people (Gordon)
Office Dogs: The Manual (Rousseau)
One Minute Cat Manager: sixty seconds to feline Shangri-la (Young)
Ollie and Nina and ... daft doggy doings! (Sullivan)
No walks? No worries! Maintaining wellbeing in dogs on restricted exercise (Ryan & Zulch)
Partners – Everyday working dogs being heroes every day (Walton)

Puppy called Wolfie – a passion for free will teaching (Gregory)
Smellorama – nose games for dogs (Theby)
Supposedly enlightened person's guide to raising a dog (Young & Tenzin-Dolma)
Swim to recovery: canine hydrotherapy healing – Gentle Dog Care (Wong)
Tale of two horses – a passion for free will teaching (Gregory)
Tara – the terrier who sailed around the world (Forrester)
Truth about Wolves and Dogs, The: dispelling the myths of dog training (Shelbourne)
Unleashing the healing power of animals: True stories about therapy animals – and what they do for us (Preece-Kelly)
Waggy Tails & Wheelchairs (Epp)
Walking the dog: motorway walks for drivers & dogs revised edition (Rees)
When man meets dog – what a difference a dog makes (Blazina)
Wildlife photography from the edge (Williams)
Winston ... the dog who changed my life (Klute)
Wonderful walks from dog-friendly campsites throughout the UK (Chelmicka)
Worzel Wooface: For the love of Worzel (Pickles)
Worzel Wooface: The quite very actual adventures of (Pickles)
Worzel Wooface: The quite very actual Terribibble Twos (Pickles)
Worzel Wooface: Three quite very actual cheers for (Pickles)
You and Your Border Terrier – The Essential Guide (Alderton)
You and Your Cockapoo – The Essential Guide (Alderton)
Your dog and you – understanding the canine psyche (Garratt)

Hubble & Hattie Kids!
Fierce Grey Mouse (Bourgonje)
Indigo Warrios: The Adventure Begins! (Moore)
Lucky, Lucky Leaf, The: A Horace & Nim story (Bourgonje & Hoskins)
Little house that didn't have a home, The (Sullivan & Burke)
Lily and the Little Lost Doggie, The Adventures of (Hamilton)
Wandering Wildebeest, The (Coleman & Slater)
Worzel goes for a walk! Will you come too? (Pickles & Bourgonje)
Worzel says hello! Will you be my friend? (Pickles & Bourgonje)

www.hubbleandhattie.com

Disclaimer
Please note that no dog was deliberately frightened during photographic sessions. The images used to depict dogs feeling uncomfortable about specific situations were taken opportunistically whilst the dogs were engaging with their surroundings.

First published in July 2015 by Veloce Publishing Limited, Veloce House, Parkway Farm Business Park, Middle Farm Way, Poundbury, Dorchester, Dorset, DT1 3AR, England. Fax 01305 250479/e-mail info@hubbleandhattie.com/web www.hubbleandhattie.com. This edition published March 2019.
ISBN: 978-1-787115-06-4 UPC: 6-36847-01506-0. © Helen Zulch, Daniel Mills and Peter Baumber and Veloce Publishing Ltd 2015 & 2019
Readers with ideas for books about animals, or animal-related topics, are invited to write to the editorial director of Veloce Publishing at the above address. British Library Cataloguing in Publication Data - A catalogue record for this book is available from the British Library. Typesetting, design and page make-up all by Veloce Publishing Ltd on Apple Mac. Printed and bound by CPI. Group (UK) Ltd, Croydon CR0 4YY

Contents

Foreword
by Victoria Stilwell

Today's domestic dogs stand alone in their ability to adapt, develop and enjoy complex and co-dependent relationships with us. As someone who has dedicated my life to making the world a better place for them, I am constantly amazed at their seemingly unending ability to amaze, surprise, challenge and enrich our lives.

My primary role as a dog trainer and behaviour expert is to help people build healthier bonds with their dogs, by better understanding how their dogs perceive the world around them. It has become rather a cliché to say that trainers actually teach people more than they teach dogs, but it is somewhat true. While dogs do need to be given the tools to learn how to be successful in this domestic world we've created, the bulk of the problems I see as a trainer stem from issues that could have been avoided if people had access to really good advice. Knowledge is empowering, but with so much conflicting information in the dog world nowadays, it can be hard to know where to turn.

Thankfully in *Helping Minds Meet*, Helen Zulch and Daniel Mills provide the information and knowledge every dog lover needs to truly understand their dog, think about the relationship they have together, and recognise how their own behaviour can enhance the bond they share.

Dogs are not born with an innate awareness that chewing a slipper is 'bad', or urinating on a rug is wrong. As their caregivers, we have to be the ones to teach them what is appropriate, and guide them away from what is not. This is not achieved by shouting or administering harsh punishment, which only serves to increase stress and destabilise the bond between dogs and humans, but instead is accomplished by taking time to understand why the behaviour happened, and to work in a positive way to modify or change it.

Helping Minds Meet will not only help you teach your dog valuable life skills, it will also make you think about how your dog might perceive your own behaviour. So much of our human language, for example, gets lost in translation. We use gestures that can be perceived as threatening, even when that is not our intention. We tend to be confusing or have high expectations that our dogs cannot meet unless we give them the tools to do so first. Zulch and Mills help you manage those expectations by providing worksheets that make it easy for you to observe, record and think about what your dog needs. It is indeed rare that a book provides so much great information while also encouraging readers to take an active part in thinking how they might deliver it.

Dogs are intelligent, sentient beings who enjoy being with us while still retaining their unique 'dogdom'. No matter how much we try to domesticate them, we must never forget that they are a highly evolved species which still retains the traits and characteristics of a predator, whose basic need for safety is paramount. Dogs do not come into our homes wanting to be pack leader, as some people might have you believe, but they will naturally gravitate towards something that makes them feel good, and protect themselves – or things that are important for their safety and survival – if needs be. This can sometimes manifest itself in behaviours that do not fit well into our human world. A support system is therefore needed to help with these problems should they arise, or even prevent them occurring in the first place.

Helping Minds Meet is more than just a book – it is a vital resource that will guide you through the intricacies of canine language and behaviour, and help you create a harmonious relationship based on consistency and trust. Teaching dogs to be successful in a domestic world is not about domination or intimidation; it is about cooperation, motivation, encouragement, confidence, and having fun. This is the recipe that ensures a successful human/canine partnership ... and a happy life together.

Dedication
To our parents and partners in life, who have taught us many of the lessons in this book through the love they have shown us.
>Thank you.

Acknowledgements
We are extremely grateful to all of the dogs and their owners who participated in the photo shoots for this book. It takes time and dedication to get the right image, but we believe that they enjoyed the process as much as we did. We hope their involvement has helped this book to connect with dog-owners everywhere.

Introduction
Our relationship with dogs has a long and varied history. However, despite his adaptability and seemingly endless willingness to try to please, the dog remains in many ways essentially the same creature domesticated thousands of years ago. In contrast, our beliefs about dogs, and our attitudes towards them have shifted, and as we integrate dogs further as members of the family, we can easily lose sight of fundamental differences between us. These differences can enrich us, but also frustrate us if we do not pause and think about things from our dog's perspective.

Dog ownership can confer physical and psychological benefits such as the increased exercise and social contact gained from going for a walk. Both owners and dogs usually enjoy interesting walks in open spaces.

The closer relationship which many people now share with their dog, together with the stresses of modern life, mean that more demands are placed on dogs to conform to specific expectations. This is particularly true of pet dogs in homes; the focus of this book. However, certain truths hold for ALL dogs, be they assistance dogs, military working dogs, or herding dogs on a farm. These demands can lead to tensions within our relationships so, more than ever, we need to remember that a good relationship is about respecting and valuing the differences which make every relationship unique.

> ### REMEMBER!
> Having a good relationship is important to both you and your dog, and depends on how each responds to the other in both good times and bad. A good relationship needs continuous input, but this doesn't mean it has to be hard work.

This is particularly important as it is an unfortunate reality that many dogs find themselves abandoned when they fail to fit into their human households, or conform to the norms of human society. At the same time, the families who relinquish these dogs suffer from the distress that a broken relationship brings. Dog-human relationships should benefit both parties: in other words, dogs should be good for people, AND people should be good for dogs.

This book aims to assist owners to better understand that, despite the inherent differences dictated by our species, we can promote one another's wellbeing and enjoy fulfilling relationships.

> ### REMEMBER!
> At the heart of a trusting relationship is the recognition that all parties usually act with good intentions, but that doesn't mean we always get it right, or that things are always as we would like.
> When challenges arise it is important that we communicate to understand each other better, rather than simply make assumptions about the other's intention.

The work you put into building a strong relationship with your dog will be rewarded as you share activities in a companionable manner.

A deeper understanding and appreciation of where the differences between us lie, means we can turn potential conflict into new opportunities: whether this is through compromise or thinking creatively so that we can find solutions where everyone wins. Our differences should add richness to our lives, rather than seem like difficulties that have to be surmounted. If you are already experiencing significant difficulties in your relationship with your dog, and are perhaps losing faith or confidence in your interactions, then we urge you to seek professional advice, rather than depend solely on this book. We all learn all the time, and hence have the capacity to change the way we behave and the relationships we have: we simply need to find the right way of supporting relationships to enable us to cope and succeed.

REMEMBER!
Key to building a successful relationship with our dogs is to celebrate and enjoy our differences when they make us happy, while respectfully managing situations when these differences could lead to conflict. This is what is at the heart of this book.

This is not a training book, but rather a guide to help you think more about what you do around your dog, and how it affects his behaviour. As relationships are dynamic and therefore always changing, we trust that at different times in your life with your dog

Watching our dogs enjoy something as simple as joyfully bounding through a summer meadow enriches our lives.

you will find different points of interest within this book. We hope that, regardless of your level of experience with dogs, each chapter will help you to reflect so that you can enrich your relationship to the benefit of you both.

Please note that, for ease of reading, this book refers to the dog as male throughout (unless a gender-specific behaviour is under discussion): however, female is also implied at all times.

A successful partnership often works not because roles are evenly split, but because each member has their own strengths which complement those of the other. While dogs and humans share a common home and can form remarkable partnerships, we have evolved in different environments with different abilities, and so need to look at how these differences can translate into the strengths that will make the relationship successful. This idea of complementary strengths doesn't just refer to different skills, such as the dog's ability to use his great sense of smell to help handlers find contraband: it might be that one individual encourages the other to be more thoughtful and therefore make better decisions as a result. Ultimately, this enriches the lives of both partners.

It is easy for the differences between people and their dogs to become a source of tension between them, but if we accept that we have different needs and we manage these with careful and imaginative consideration, we can maximise the positive potential in our relationships with our dogs.

REMEMBER!
The things that make your dog happy may be different to the things that make you happy and vice versa. Individual differences are an important feature of relationships that can be used to create a stronger team. It takes planning and forethought to allow both of you to have what you want without coming into conflict, eg by setting boundaries and rules for you and your dog or ensuring that certain behaviours can be controlled by providing appropriate training.

A dog's priorities are not the same as a human's

While all animals prioritise things like staying alive, dogs and humans will, in many situations, have different priorities. For example, dogs have a phenomenal sense of smell, and seeing your dog use his nose to trace a scent that we cannot perceive can be a truly enjoyable experience. Such differences reflect the joys of the partnership we can form with dogs, but they usually come with a potential downside, too: dogs may be disturbed or distracted by

Dogs, especially when young, like to chew. Providing for this need (by giving him safe chew items or chew toys appropriate for him) will reduce the risk of conflict caused by him chewing on your precious possessions.

changes in the chemical environment that might go unnoticed by us. For example, they may react to the smell of air fresheners, or a bitch in season nearby. Some chemicals in the environment – such as those in the latter case – can have a powerful effect on a dog's mood, so it is always important to bear in mind that, in a situation where your dog's behaviour changes, and you can't see a reason for it, an odour we can't smell, or one which seems irrelevant to us, may be influencing him.

TOP TIP
Give your dog time to investigate the odours in his environment. It's

Sometimes it's the simple things, like stopping on a walk to watch livestock, which give your dog pleasure. As long as it is safe to do so and he is behaving appropriately towards whatever is holding his attention, allowing him to take time to look is beneficial to him, and likely to teach you more about his character, too.

These dogs are enjoying sniffing together on a walk, and so are meeting a variety of their needs, including taking in important information about their environment and sharing relaxed companionship.

important for dogs to sniff around in a new environment as well as a familiar environment where the scent picture may have changed. Respect this need. In addition, if he tells you through his behaviour that he is feeling uncomfortable, respect that and respond appropriately to help him feel better (see chapter 5), even if you can't identify a reason.

This ability to investigate odours is important, whether he is on- or off-lead, so it is useful when he is on-lead to teach him cues which mean 'you can go sniff', versus 'no sniffing now' (see Appendix 1) to reduce potential conflict arising from your different priorities: when you need to get from A to B quickly, for example, and he wants to read the pee-mail!

Most dogs don't care about getting wet and muddy, and they don't recognise that the cleanliness of your new carpet is important to you. However, this potential source of conflict can be avoided by implementing certain rules or routines that allow him to get mucky, but ensure that he accepts being cleaned before he comes into the house. This is a much fairer compromise than simply depriving him of outdoor pleasures.

> ### Remember!
> He may also need to leave his own scent mark in the environment so allow him to do so, as long as it is appropriate from a human perspective!

Another reason that our priorities differ is that we provide for or control many of the basic biological needs of a dog (eg food, shelter, breeding via neutering, etc), whereas we may have to work towards achieving these for ourselves. A key way that this

It is normal for dogs to sniff things in the environment that we may find disgusting, such as another dog's faeces. Making a fuss over this could create a problem where none need exist. Of course, if it appears that he is about to roll in or eat something that you would rather he did not, you can calmly call him away.

Urine marking is a normal behaviour in both male and female dogs.

Remember!

It's important to allow your dog to *be* a dog, and it's okay to simply give him free choice sometimes, when it is safe to do so. Enjoy what he chooses to do (and learn from it) rather than being upset or disgusted. If he does engage in things you don't like or which you worry may be detrimental to him, don't think that complete restriction or total avoidance of situations or areas is the only solution. Creative planning for situations (such as using distraction techniques, or teaching him alternative responses) can enable you to enjoy outings in risky areas while keeping you both safe and happy.

If you would rather your dog did not engage in a certain activity, such as splashing through a muddy puddle, you can put him on a lead, or use a toy to distract him from the puddle as you move him quickly past it.

may influence our priorities in different ways is that he may see interaction with you as a priority, whereas you need to prioritise your job and other activities that take you away from him. Some dogs may find the separation caused by your absence stressful*, or they may simply find being alone uninteresting. You therefore need to ensure that your dog does not suffer as a consequence of being alone, and has appropriate care during the day, including activities to engage him, and that, when you are home, you give him the interaction opportunities he enjoys.

TOP TIP
It is useful to teach your dog a signal that means 'I really can't give you attention now' so that he knows that sometimes he can come to you for fuss on his own terms, but sometimes he has to respect that you are busy. If you can convey this clearly, politely and consistently, he will learn that, at those times, he needs to amuse himself. This will avoid him becoming upset or bothering you with annoying attention-seeking behaviour (see the photos on pages 50 & 51 for more info on this).

Key psychological needs
As well as social interaction, safety and security are amongst the most important psychological needs your dog has, both within the home and when out in society. Most people never need to consider what it may be like to frequently feel frightened or anxious, as most of us live in relatively safe societies, and we understand what is happening around us. However, this is not necessarily the reality for dogs. For this reason it is important that your dog has a safe haven in your home (see Appendix 2) where he is in control, and knows he will come to no harm. Everyone should respect this space, and allow him to access it whenever he wants, and not bother him when he is there.

　　When out and about, and especially in social situations, we need to ensure (and convey to him by being supportive) that no harm will come to our dog when he is with us. In reality, this means ensuring we are consistently dependable, so that he can have confidence in us in all situations, even if we don't understand the reason for his behaviour, or if it embarrasses or frustrates us (for example, if he growls at another dog).

　　We may not always love what our dog does, but we should always love him and support him, even when he makes a mistake. Losing our temper is not only counterproductive, but deprives our dog of the social support he needs to become a confident individual. This is especially true when his behaviour is motivated by fear or anxiety – for example, if he is trying to avoid something. This does not, of course, mean that we simply accept and live with

* If your dog is noticeably distressed when separated from you please seek help from an appropriately qualified professional.

Every dog should have a safe haven in his home. This may be a bed, a basket or a crate. The key point is that it is a place where he feels safe and secure. See Appendix 2 for details on how to establish a safe haven.

behaviours that are inappropriate or difficult to manage, but we need to put a long-term plan in place to change his behaviour whilst supporting him in the short-term.

Enrichment is a personal thing
Another key psychological need we share with dogs is to mentally engage with things: ie to explore the environment and seek out

REMEMBER!
Just as we meet a dog's essential physical needs, certain psychological needs must also be met to maintain good well-being.

information about it through play and investigation. As we have the ability to control aspects of our world – to travel, watch television, read a book – we may not appreciate that we are naturally meeting this need on a daily basis, but that most dogs do not have the

Allowing off-lead exploration is (for most dogs) one of the best ways of meeting their need for investigation and interaction. Therefore, learning to be safe and controllable during off-lead walks is a key skill for dogs to master.

new activities or devices, and watching his response. This can be enriching for you, too, especially as you discover hidden talents in your dog. Do bear in mind, though, that many of the newer dog games on the market require you to do a bit of training with your dog for him to understand them, so set aside some time for proper introductions.

If you're just starting out and unsure about what your dog may enjoy (and don't be bound by breed stereotypes), buy or make

opportunity to do the same. It is therefore our responsibility as owners to meet this need in our dogs through enriching their environment.

We all value different things in life, which means that some things are more interesting to certain individuals than others, and while there are some basic principles to enrichment – such as ensuring the recipient is capable of using a specific device, and balancing predictability and control (see below), just because a manufacturer claims it's produced a great toy, it does not mean that your dog will necessarily agree.

You may have a good idea of what your dog enjoys, but don't be afraid of offering him new opportunities by introducing

Interactive food toys can provide dogs with mental challenge and enrichment. Spending time teaching your dog how to use these toys is an enjoyable way to build your relationship, too.

REMEMBER!

If your dog is simply not interested in a toy or activity, it's useless, and not enrichment – regardless of what it says on the packaging. Likewise, it does not have to say 'toy' to be a plaything, though be careful about drawing clear distinctions between play items and similar non-play objects in the home: eg it's not a good idea to give an old T-shirt or slippers as a toy, as you want to avoid the risk of your dog believing that all such items are toys.

Helping minds meet

a selection of inexpensive (but safe) toys, and offer them all to your dog to see which attract his interest most.

Tips for providing enrichment

- Provide meaningful enrichment. Look at the toys and other playthings your dog has or has had, and note which ones are used and when, and which are not. Are there any common themes linking the different items, eg: are the chew toys all but destroyed but food dispensers ignored? Does he prefer a certain size of toy, or toys made of certain materials? Consider what this tells you about your dog's preferences, and seek to provide variations on this theme.

- Rotate the provision of toys, etc. Dogs can quickly get used to things, so it's helpful to have a cupboard of toys that allow you to swap around on different days. This may be particularly valuable for any toys that your dog liked initially, but then seemed to go off. Pick up some of the toys that your dog has left lying around and put them away for another day, replacing

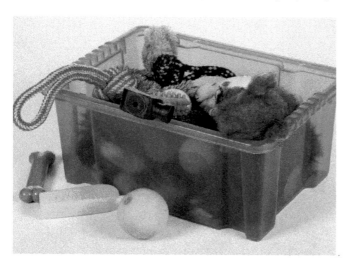

Providing your dog with a toy box containing a variety of toys enables him to choose which he feels like playing with at any given time. This box would be better for having some toys taken out (and stored in a cupboard for later), so that the range presented to him can be rotated to keep his interest. It also appears that some of these toys are never played with (they look brand new!), so therefore, to this dog, they are not acting as enrichment. Stock your dog's toy box with the type of toys he prefers.

them with different ones from the store. However, if your dog has an item that he carries around like a 'security blanket,' pop this back into his safe haven, rather than put it away.

- Balance predictability with control in the home environment. Good welfare depends on having the right amount of control over your environment, balanced against a certain level of unpredictable but largely pleasant surprises. This keeps the mind happy and healthy. Too predictable a routine becomes boring, and too unpredictable a routine unsettling.

  Try to generate and teach your dog signals that reliably predict certain routines – such as he always gets groomed on a specific mat, so when the mat comes out he knows what to expect. In addition, teach him that there are appropriate behaviours that will pay off when he predicts something may happen, like sitting by the door calmly when you indicate it's time to go for a walk. He can then potentially use these polite learnt responses to communicate what he wants, too (such as some fuss from you), and give you ideas for what to do and when.

- Occasionally, give your dog a treat for simply being a great companion. Nice surprises can help to strengthen the bond between the two of you, but it is important that these remain surprises and not an expected routine. So, occasional and varied, not predictable and indulgent (and they don't have to be food!).

We are all individuals

Quite apart from the differences which exist between dogs and humans, everyone is an individual; the product of their genetics and early environment. This means there are certain things about our dog's character that we must simply learn to accept. For example, you and your dog may have very different friends, and we should respect this rather than force friendships where they do not seem to occur naturally. While there are procedures (beyond the scope of this book) which can be implemented with professional

Remember!

Marketing people design toys to be as much or even more attractive to the buyer (ie the person) than the user (the dog). Many toys are red, for example, because this shade is attractive to us, but dogs don't easily see this colour in the environment, so before you buy that new toy, ask yourself how much it appeals to you rather than your dog, and go for those that seem to fit with your dog's personality rather than yours.

Performing specific activities in a predictable way – for example, always grooming him on a specific mat – enables your dog to predict what will happen next. This helps to reduce the stress which can accompany a very unpredictable lifestyle.

support to help encourage affectionate relationships, we must never try to force together two individuals. If your dog seems to take a dislike to someone or another animal – which might be due to a previous experience (or lack of experience) and related associations – in the first instance it is usually preferable to simply accept that they don't get on, even if the person wants to be with your dog. Keep your dog away from this individual by making alternative arrangements for when you meet up.

Individual differences and their associated unpredictability are partly the reason that we form close relationships, which means we will continue to be surprised by our dogs from time to time, and our relationship will not always be as smooth as we might like, as we will not always get what we want from him. Developing a frame of mind that recognises and values this is as much the secret of a happy relationship as is what we do for each other.

Respect that your dog may want to avoid an interaction with another individual: for example, as in this image, by veering away and showing tension in his face. Just as we don't immediately like everyone we meet, neither will your dog, and forcing him into an interaction he wishes to avoid may mean he feels the need to express his opinion more overtly – by growling or snapping.

WORKSHEET 1

List your most important activities, ie things that must get done each week, and whether or not you can engage your dog in these in any way

Activity   *Possible ways you can engage your dog*

. . . .

. . . .

. . . .

Are there skills your dog may need to learn to be able to engage with you fully in these? (eg he could go with me to the park when I meet friends for tea if he could settle quietly at my feet)

. . . .

. . . .

. . . .

List the things you know your dog enjoys, and whether or not you can share any activity with him

Activity   *Can I enjoy this activity with him? If so how?*

. . . .

. . . .

. . . .

When my dog can't engage with me in an activity, can I direct him to an activity that I know he will enjoy? (eg when I need to go shopping can I provide him with a chew toy to keep him amused while I am gone)?

. . . .

. . . .

. . . .

2 There are limits to what we should expect from one another

In an ideal world everyone would live in harmony, and it would be possible to meet everyone's needs and fulfil everyone's desires at all times. In reality, we constantly have to compromise: one or other member of a partnership has to sacrifice what they want at any given time, or both agree on an alternative. For example, there will be times when all that you want to do after a long, stressful day is put your feet up in front of the television – but your dog just as desperately wants a walk and food before settling for the evening. Likewise, from your dog's perspective, there will frequently be times when he would like to have the freedom to explore, interact, and run around freely when you would rather he stayed close and calm.

It is important to balance the times when each member of the partnership gets their way, because if one side always wins out there is a risk that the welfare of the other may be compromised, or that the one who constantly wins forms expectations that this will always be the case – which may be a dangerous assumption. For example, if your dog has learnt that he can always have his pick of the best seat in the house and refuse to move, it could result in him defending this learnt right if a visitor tries to insist that he move: one reason why setting boundaries and teaching dogs to comply with polite requests is so important.

Learning to recognise and, where possible, anticipate where our needs might conflict with our dog's can help us find the best way to help him accept compromise, and it is fairest to do this by identifying clear boundaries and rules at an early stage, as these will ensure everyone's wellbeing.

Boundaries and rules
Before we can expect our dog to behave in accordance with our wishes, we have to ensure that he knows what these are, and understands them. It is important for the whole family to have a common set of rules and boundaries which your dog can then be taught. So, for example, one rule may be that he may not climb on the furniture; another that he should greet people without jumping

There will be times when either you or your dog is reluctant to engage in a specific activity at a specific time. As long as, in general, both parties' needs are being met, it's fine to simply accept this and engage again at a time or in a manner which suits you both. If, however, your needs are so at odds with one another that it is a common occurrence, and it is impacting on the welfare of one or both of you, it is important to take stock and consider what changes should be made so that expectations are met and needs fulfilled in the longer term.

Most dogs enjoy digging (especially when they are young), but most owners don't like their lawns or flower beds being destroyed! These opposing needs can be met through creative planning: for example by creating a digging pit in the garden. Demarcate the digging area using logs or stones to create a visual border, loosen the soil in the centre and half-bury chews, treats or toys to encourage him to focus his attention on this spot.

up, and a third that he needs to listen and respond appropriately when he's asked to do something. Rules for the family are that each member ensures the house rules are consistently implemented in a manner which their dog can understand, whilst at the same time being certain that –

- He can still get the fuss he desires on the floor.

- People will acknowledge him and give him the attention he craves if he manages to keep all four paws on the floor.

- People will consider carefully before asking him to do something which may be at odds with what *he* wants, and also acknowledge this fact by rewarding him when he complies.

- His needs are met adequately on an ongoing basis.

If it's likely that you will sometimes want your dog on the furniture and at others you may not, it's better to teach him the rule that he can only get on the furniture when invited, rather than risk conflict. This is what we mean when we talk of finding an approach that keeps everyone happy. However, we have to recognise that this puts the responsibility on us to teach our dog this rule in a compassionate way.

If one of your rules is that your dog is not allowed onto the furniture, he will find it easier to comply if you provide him with a comfortable bed on the floor near to where you are sitting.

HOUSE RULES

FURNITURE:

- Dogs may get on conservatory furniture at any time – even if they are wet or dirty

- Dogs may NEVER be allowed on living room furniture

- Dogs may get onto beds only when invited

GATES, DOORS ETC:

- Dogs must wait at every door or gate to be invited to go through.

VISITORS:

- Dogs must sit to greet visitors

Writing a list of house rules and placing them in a prominent place will make it easier for the family to consistently apply them.

The concept of 'don't do something' is difficult for dogs to grasp, and it is much better for them to learn what TO do in every situation. Teaching your dog that if he goes to lie on his bed when you are busy and he might interfere (eg when you are cooking or eating, and he might be tempted to beg for food) he will earn treats, means that he can learn to make the right choice in these situations. Over time, the treats can be faded out, and this well-behaved behaviour will become a self-rewarding habit.

Tips for consistency

- The whole family must agree to and be prepared to implement the rules. Write them down and display them in a prominent place.

- Rules must be implemented consistently, every single time by every single person – just one titbit fed from the table can set back your training by weeks. This takes self-discipline, especially on the part of children and soft-hearted visitors!

- Think through strategies in advance so that these can be implemented to enable your dog to make the right decision – that is, to follow the rule. For example, if you don't want your dog to beg at the table at mealtimes, place a dog bed a little distance from the table, and toss tasty dog treats onto it while you are eating your meal. Over time you will see your dog automatically go to his bed when you sit down to eat, and eventually you can fade out the rewards, and simply tell him how good he has been at the end of the meal.

- Manage the situation to prevent unwanted learning if you think a situation will arise where a mistake might occur – for example, if a toddler who is likely to drop food under the table is coming for a meal. Keep your dog away from the table with a desired food toy or chew to occupy him.

It is also important to appreciate that dogs won't know instinctively when rules change. For example, if they are allowed on the furniture in your house, they may well expect to be allowed to get on the furniture when visiting a friend's house, or when you are in holiday accommodation.

Tips for adapting rules

- Ensure that your rules will easily translate to different environments and situations your dog may encounter. This is where teaching him that the boundary can only be crossed upon specific instruction (for example, when he is invited onto the furniture) can be useful. This is, however, more complicated to teach and maintain, so it's important to think through all aspects of rules ahead of time.

REMEMBER!
It is much easier to do something specific than simply to stop doing something, which leaves a dog with many other choices, some of which may still be the wrong one. Always have a plan so that you can direct your dog towards what's wanted.

- Have a plan of action for every situation where he will encounter the rule change, so that it pays off for him to follow the new rule whilst ensuring that people do not get angry with him when he makes mistakes. For example, if he is no longer going to be allowed into a certain room because of the arrival of a new baby, install a stair gate across the doorway, and make it enjoyable for him to stay on the other side – by providing him with a food toy or chew every time you close him out. If installing a stair gate is not possible, place a comfy dog bed away from the doorway, and teach him to stay there whenever you go into the now-forbidden area.

REMEMBER!
For some dogs learning things like 'I mustn't jump up at the kitchen counters' takes time (even many months), so ensure the family has strategies to reinforce alternative, appropriate behaviour until it becomes habit for your dog to do this.

- The longer he has been used to the old rules, the more his behaviour will have become habit. We all know how difficult it is to change habits, so if your dog has learnt one set of rules that, for unforeseen reasons, have to change, it is important to appreciate that new rules need to be taught with kindness and compassion, and also recognise that errors will be an inevitable part of the learning process. Support success and ignore or gently redirect failure.

Recognise when your dog needs help

Dogs are not spiteful, and nor do they 'know better,' but they can be distracted and confused. When it comes to our dog's responding appropriately to requests from us, there will be times when, no matter how well he has learnt something, he may not comply as well as we would like him to. There are many reasons why this may be the case; understanding some of these can help us, firstly, to keep calm in the face of his perceived disobedience, and, secondly, to understand how to make it easier for him to follow our instructions in the future.

Although it is often natural for us to learn something in one situation and then apply it in another, this sort of extrapolation is not that easy for dogs. For dogs, especially in the early stages of learning, but also with established responses, what they learn is at least partially connected to the situation where they have experienced it. So, if you teach him to 'sit' on cue in the kitchen, it takes a little bit of practice before he realises that 'sit' also means 'put your bottom on the floor' when he hears the word in the living room, in the garden, or in the park. Over time it becomes easier

Above & right: Once your dog has learned to 'sit' in one situation (for example, in your kitchen), it is important to re-teach it in a number of different situations, until he realises that 'sit' means 'sit' anywhere, at anytime. Lots of practice means that he will then be able to respond quickly and reliably if required, even in exciting or difficult situations.

Practising controlled behaviours, such as 'lie down and stay' in a range of situations means that you will be able to manage him more easily in risky situations: preventing him from jumping out of the car, for example.

for him to generalise new behaviours to new places, but in the beginning it is really important to help him.

Tips for aiding generalisation
- Once you have taught a behaviour in one place, practise it in many different places, if necessary re-teaching it in new environments to start with.

- Make sure different people practise asking him for the behaviour, too, so he realises he can respond to requests from different people (assuming this is what you want).

- If your dog appears confused when you ask him to do

Helping minds meet

something, perhaps you are asking him in a way that may be confusing to him. For example, if your hands are normally free when you ask him to 'sit,' he may not respond as you wish if you ask him when your hands are full of shopping bags. When we interact with our dogs, our body posture provides a lot of information about what it is we want, so be aware that your dog may pay more attention to this than to actual words.

Even if you don't believe it is necessary for your dog to perfectly respond to verbal as opposed to body language cues, there are times when this is really useful: if you have your hands full (so can't give him a hand signal), and you need him to 'sit,' for example. As dogs focus more on our body language than our verbal requests, explicitly teaching him how to recognise verbal cues is important.

It is difficult for your dog to come back to you if you are calling him away from a situation where he is having fun. It is important to train for this type of situation, as immediately coming when called can be a life-saving response. It is however important to FIRST teach him to come in easy situations – when there are no distractions around (see above image). Only once he can reliably come back in these situations is it time to introduce distractions (as in the image below, where his friend is having a game). It takes many hours of practice to train a reliable recall.

Another reason our dog may get it wrong when we ask him to do something is that it is just too difficult for him – physically or mentally. For example, if he has pain in his back or legs, sitting at every kerb as we walk through the streets may be very uncomfortable. Alternatively, if he is worried by workmen at the side of the road, then asking him to lie down quietly while you chat to a friend might be just too scary. Finally, and especially when he is young, the motivation to do something else rather than follow the rules or your instructions – like trying to grab the toy from your hands instead of politely waiting for it to be given to him – may simply be too great at this time.

If you do need to call your dog away from something that is really important to him – such as exploring with a friend – make sure that you reward him with something he really values – a special toy, say. Of course, as with all behaviours, it is not necessary to reward every correct recall, but when we are asking him to do something difficult, we stack the odds for repeated successes in our favour if we teach him that responding to us really pays off. And it's fun for owners to engage in these games, too!

HELPING MINDS MEET

Tips

- Make learning as simple as you can and build up behaviours in simple steps, so you keep his learning within his ability at all times. Every time your dog feels good about something, he is likely to do a similar thing in similar circumstances, whether or not it pleases you. So, if you have a dog who finds chasing intrinsically rewarding, every time he chases, his tendency to chase in future gets stronger, and he may learn to ignore your calls too! To teach him to come when called instead of running off after moving objects, you need to teach him in a way that is easy for him not to make a mistake (see Appendix 3 for texts on teaching recall training). For example, find a safely fenced area and a willing friend. Arm yourself with his favourite treats or toys and practise first calling him away from your friend just standing still, then walking away, then jogging away and finally, if he can manage it, your friend can run away playing with a less favoured toy while he comes back to you for a really good game. Always teach the easy stuff first!

- Always make it very clear that it is worth his while to comply with your requests. Using a reward that he really values can help, so saving special food treats or toys for training sessions, especially in more challenging environments, can help him to be more motivated to listen to you than to follow his conflicting desires. With most dogs, over time, you can fade out the extrinsic motivation as he learns to derive enjoyment from working with you. However, if you know you are taking him into a situation which he may find particularly challenging (for example, if he likes to chase moving objects and you need to walk him along a cycle path), ensure that you take precautions to prevent him from being able to make a mistake (keep him on his lead), and that you have a high value reward to give him for remaining calm and paying attention to you (treats or a tug-toy as he exercises self-control and does not try to lunge at the passing bicycles).

- Take time to consider whether something physical (eg joint pain) or emotional (fear) could be impacting on your dog's ability to respond to your request whenever he doesn't seem to react as you expect. This is especially important if you are sure that he understands your request and that he should be motivated to comply (ie: you are not aware of anything in the environment that is more attractive than doing so).

- Or maybe something is just not pleasant to do right now – like sitting on cold, snowy ground, in which case, maybe we should brush it off, not insist, and just move on. Yes, ideally, we don't ignore non-compliance with our requests, but in a well-trained, generally obedient dog, a single episode of 'disobedience' that is without consequence is unlikely to undermine all our hard work. Of course, this should not become a habit, so we must learn from it to avoid repetition, by either modifying our requests in a similar situation or setting up things differently to ensure compliance. However, getting angry at that point is unlikely to achieve anything constructive.

A reality check

Finally, we also need to appreciate that, just as we have off days when we forget to do something, don't pay attention, or are finding it difficult to motivate ourselves to comply with someone's reasonable request, so, too, do our dogs. Let's accept this and treat them with sympathetic understanding, helping them to make the right choices, rather than impatiently insisting on instantaneous obedience.

WORKSHEET 2

These are the rules of our house and family (eg our dog must stay on his bed when food is being prepared)

Rules

. . . .
. . . .
. . . .
. . . .
. . . .

These are situations where rules may need to be adapted, and a plan for how to achieve this

Adapted rule
(eg when we visit family, he can't get on the sofa)

How we will teach this
(eg we will teach him that he can only get onto the furniture when invited)

. . . .
. . . .
. . . .
. . . .
. . . .
. . . .
. . . .
. . . .
.

Do seek help from a professional trainer to enable you to achieve your goals, rather than abandon one, because you cannot see a way to do it yourself.

3 We all make mistakes from time to time

As hard as we may try to get through life without upsetting people or breaking society's rules, sometimes we do. The same goes for our dogs: no matter how well they understand the rules, or how obedient they are, they will make mistakes. So on the day that you unthinkingly tell him off as he happily drops a smelly, spat-upon tennis ball on your lap as you try to relax after a busy day, don't beat yourself up. Recognise that there is a more appropriate way to respond and move on.

On the flip side, when your usually well-behaved dog becomes over-excited, forgets his manners and leaps up at your neatly dressed guest, remember that he, too, can make a mistake. And a mistake is just that, a decision made in error. It isn't deliberate; we need understanding for our mistakes, and in fact they can be important learning opportunities.

As discussed in the previous chapter, there are many things we can do to help our dogs fit into our lifestyles, and not fall foul of our rules and expectations. However, when mistakes happen, there are a few points worth remembering that can help us make things better in future.

> REMEMBER!
> One way to avoid mistakes is to be sufficiently aware of what your dog is doing to be able to intervene and direct his behaviour BEFORE he makes a mistake. For example, if you know he likes to chase joggers and you see him notice someone running in the distance, immediately focus his attention on you and, if necessary, put him on a lead – before he can set off after the runner. It's better to be able to direct the intention than have to correct the unwanted action.

Tips for helping you deal with your dog's inevitable mistakes

- Don't respond as if he is deliberately trying to annoy you, because dogs don't act out of spite. As far as we know, dogs don't have the mental capacity to deliberately decide to cause someone hurt or annoyance. Yes, they can make a deliberate choice to disobey rules or requests, but it's simply because it

serves them better to do this – they get something out of it, they don't do it to 'get back' at us. If you feel that the action is not a one-off mistake, but becoming a more consistent pattern that needs changing, consider the information elsewhere in this text, or seek professional help if necessary.

> REMEMBER!
> Respond; don't react!

- There is no value in trying to make a dog feel guilty, or responding as if he knows what he's done is wrong. The 'guilty look' our dogs appear to have is, in fact, their way of trying to appease us, and defuse a situation that they perceive as threatening (see Chapter 4). These are the same behaviours they show to one another to avoid or stop a dispute, or make up after one.

- Get your dog to focus on something else, and once he has focused away from the situation that is triggering the response you deem inappropriate, you can more easily direct him to do something appropriate. Trying to direct dogs when you do not have enough of their attention to enable them to respond to you is usually futile. Methods for gaining your dog's attention include the correct use of his name (see chapter 5, page 52 for information on teaching him and using his name), or a specific, attention-getting cue.

Alternatively, other simple methods which might work, depending on context, include –

- Running in the opposite direction, and acting as if you are having a great time.

- Giving attention to another dog.

- Moving directly into his line of vision, or holding a food treat in front of his nose to encourage him to turn to you and focus on you.

continued page 28

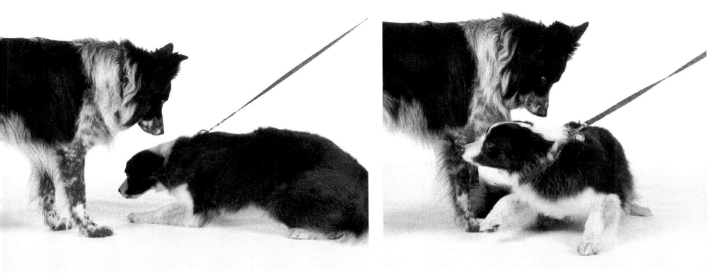

The smaller Collie in these images is displaying signs which people typically interpret as 'guilt' – body held close to the ground, hindquarters and tail tucked under, head turned away, ears pulled back, lip lick (first image) and lips elongated (second image). However, rather than indicating that a dog knows that he has done wrong, these are, in fact, signals aimed at letting another individual know that he is trying to avoid conflict, or re-establish a positive relationship after conflict. Therefore, getting angry with him at this time is counterproductive. Note: if dogs do meet on-lead, ideally, the leads should be kept as loose as possible (unlike the taught lead in this image) to allow the dogs full opportunity to move as they wish.

This Collie is behaving calmly even though he is interested in the sheep. As there is a fence between them, his owner knows she can call him back in an emergency, and the sheep are at a distance, so she has decided that the walk can continue without putting him on-lead. However, it is still important that his behaviour is monitored closely to prevent him from trying to chase the sheep.

In this situation, the owner has judged that the horse is too close for her dog to be able to ignore, so she has focused his attention on her to calmly but quickly move him past. Once your dog has learnt that focusing on you is enjoyable, you will no longer need (in most situations) to use treats or a toy as a distraction, but simply talk to him and keep his attention for as long as you need to, giving him a release cue ('go play' for example) to let him take himself off and engage with the environment again when it's appropriate.

Simply giving your dog time to stand and look at a situation which is worrying him is often sufficient to allow him to cope; you can then continue with your walk or interaction. If, on the other hand, as he stands and looks, he begins to get more upset or over-excited, it's advisable to calmly and gently move him away from the situation.

Helping minds meet

- Think about what is likely to happen next –
  + If you remain in the situation can he continue to behave appropriately, or is the situation too difficult for him to make the right choices?

- If the former, keep a close eye on him to check that he continues to cope appropriately. It may be necessary to put him back on his lead in some situations.

- If the latter, calmly and quietly move him away from the situation.

- If you need to move him away, but know that it's a situation he will need to confront again, commit to making a longer term plan to deal with the situation. This will ensure he will learn to make correct decisions and behave appropriately, rather than run the risk of making the same mistakes again – repeated mistakes can become habits.

If you don't indicate to your dog what you need him to do, don't be surprised if he makes his own decision. In this case, he is not making a mistake, but simply making a decision which he judges best in the absence of alternative guidance. It can be hard for our dogs to know what is right according to human rules, so it's up to us to direct them in how to be well behaved. This is the best way to avoid him getting himself into trouble – and the best way to ensure we can take him places and trust him to be a credit to us, and an ambassador for companion dogs everywhere.

Leaving most dogs unattended in a room with tasty food available at their level is likely to be one temptation too many. Some dogs will even try to steal food when people are around, which can lead to reprimands and conflict. It is much better to teach your dog that when there is food in the vicinity, it pays off for him to go to his bed, where good things are likely to come to him. In this way, dogs choose to be well behaved, rather than having to be constantly directed, which makes life simpler and more pleasant for all.

WORKSHEET 3

Situations where mistakes commonly happen (eg although he generally has a good recall, when other dogs are around he won't come back to me)

1. . . . .

2. . . . .

3. . . . .

4. . . . .

Things I can do to prevent these from happening (eg in the short-term, when I know other dogs will be around and I don't want him to run off, I will keep him on the lead)

1. . . . .

2. . . . .

3. . . . .

4. . . . .

Skills I need to teach my dog so that he is less likely to make the same mistake next time (eg I will ask friends with friendly and well behaved dogs to help me set up a training plan to teach him to come to me, even when he is around other dogs)

1. . . . .

2. . . . .

3. . . . .

4. . . . .

4 We don't speak the same language

At the heart of any good relationship is an ability to communicate well – to hear as well as be heard. Dogs and humans differ both in the way they naturally receive information – hear – as well as in the way they transmit it – try to be heard. By understanding where these differences lie, we can help our dogs to better hear what we are saying to them, whilst at the same time honing our skills for listening to them.

In this chapter we focus on our understanding of dogs and their intentions, and in the next one we focus on how our dogs see us.

The starting point for dogs and humans

A major reason for these differences in communication styles lies in the fact that we place a different emphasis on the information we receive through our senses, and even prioritise different elements within a given sensory channel. Therefore, we experience the world in very different ways.

People, in general, use predominantly visual information to examine and comprehend the world around us, and sound for detailed social communication, whilst dogs seem to use odour for closer inspection and detail, with sights and sounds providing information from afar. However, even when we consider the visual world, it appears that we have different priorities. While humans pay most attention to shape when classifying objects, dogs may pay more attention to size. So, to a human a ball is a round object, but to a dog it seems to be 'something about this size that I fetch'. There is a lot we still don't know about this sort of world view, but it clearly has important implications when we consider training dogs, and may explain where they might make mistakes.

Consider how important each of the following channels of communication are to you, in a specific exchange, and then try to think about this from your dog's perspective –

Dogs and people see the world in very different ways. The first image shows how we are likely to view a scene at human height, whereas the second gives an approximation of a dog's eye view. (This image does not fully account for all the differences in dog and human vision, but simply illustrates some of the differences in the visual information available to each of us in the same context).

Dogs and people have different sensory capacities, and also place a different emphasis on the sensory information they get about their world. On this walk, a person may admire the colour and pattern of the onion crop, while the dog would rather have a good sniff! Allowing dogs (where appropriate) to investigate their world in the manner they prefer will benefit their well-being.

* Visual signals – body posture, gesture and facial expression.
* Auditory signals – vocalisations.
* Chemical signals –scents and pheromones.
* Tactile signals – different forms of physical contact.

     In many situations dogs will tend to favour the first and third channels and humans rely, at least consciously, primarily on the second (though the fourth will be construed as very significant if it occurs), so it's easy to see how miscommunication can occur.

     In addition to being able to listen and convey messages effectively across the species barrier, we also need to understand that our different species' histories have primed us to naturally use and accept forms of interaction which the other species may not find inherently appropriate. For example, humans greet face-to-

(Right) The dog in the first image is uncomfortable with her owner hugging her – note the pulled back ears, turned head and averted gaze – while the second dog has learnt to enjoy this contact as he actively leans into the embrace. Never assume a dog will enjoy this kind of contact, especially from someone he doesn't know well. In any interaction, keep an eye on his body language to gauge whether his is enjoying or wants to avoid or escape the situation.

face, with direct eye contact, and may hug one another, whereas dogs are much more indirect in their initial approach, and so likely to find each one of the usual human gestures – let alone all of them combined – threatening, especially from a stranger. Many dogs may come to accept and even enjoy such interactions with their owners, but for some this may simply be a step too far because of the strength of their species bias. This isn't anyone's fault, but we should respect our differences, and manage things accordingly.

REMEMBER!

Consider how our dogs may feel about gestures that we consider to be friendly towards each other, and those which dogs use to introduce themselves. Try to avoid misunderstandings, especially with new dogs, by being sensitive to the species greeting and interaction biases that exist.

It is always more appropriate to greet a dog by reaching to touch him under the chin or to the side of his neck.

Hints and tips for better greetings

• Allow your dog to take the initiative when greeting strangers, because moving into a dog's space, or reaching out to him – especially over his head – can be regarded as threatening. Ask the stranger to remain passive whilst your dog moves towards them, and let him sniff at their hand first before they initiate any contact (and then only if he appears to want it). This should involve touching him gently but firmly on the side of the shoulder or neck without reaching over him. If he chooses to move away, let him: he's not saying he dislikes the person, but rather that he is not yet comfortable with close or physical contact – just as we may not be comfortable hugging someone at our first meeting.

• Rolling over is not necessarily a sign that a dog wants a belly rub. Even if your dog solicits belly rubs from you, and you believe that he enjoys being hugged by you, he is unlikely to enjoy such intimate contact from people he doesn't know so well. Dogs who roll over on their backs in the presence of strangers are more likely showing a signal that they are trying to avoid further interaction. This is understandably difficult to interpret as, superficially, the two appear quite similar. Look for subtle signs such as a relaxed tail and mouth as an indication that he is comfortable, and a tucked tail and tension in his lips when he is not. If in doubt, play it safe and ask people to back away from him and give him a bit of space. If your dog seems to enjoy having his belly rubbed by you, recognise that this is a special privilege, and not something given to everyone.

• Jumping up towards the face is a natural dog greeting that needs to be substituted with something more acceptable, as jumping up is usually considered inappropriate, if not scary. To help your dog control this urge, and prevent him from becoming confused about what is and is not allowed (if some people tell him off for jumping up, whilst others encourage him), you need to teach him how you would like him to greet others. We suggest that this should be to sit politely when he is greeted. Confusion about what to do when meeting new people can quickly lead some dogs to show other potentially inappropriate responses, such as barking.

REMEMBER!

Dogs often appear very smart because they are extremely sensitive to the signals we give out, and respond to these so quickly. This can lead to some believing that dogs think like us, and can give rise to misconceptions about what dogs are 'saying,' or why they are doing something.

If you don't know a dog well it is also important to allow him to approach you to greet you rather than you moving into his space. This person has invited over the dog to greet her, and he has walked up to initiate the greeting ...

... and she gently strokes him on his shoulder. You can see the dog's relaxed body posture and facial expression, which demonstrates how comfortable he feels in this situation.

Dogs don't only roll onto their backs when they are seeking attention, they also do so to express the desire for another to move away from them. This dog's raised head with slightly widened eyes, and his tucked tail, tell you that he is not relaxed and seeking attention, but rather that the person approaching needs to give him more space.

Jumping up to greet people is normal dog behaviour, but this is not appropriate as it could injure or frighten a person. Teaching your dog to sit calmly in his bed when someone comes to the door encourages better behaviour, as he can then enjoy a polite greeting once the visitor is settled in, if this is appropriate.

Common ways dogs express themselves

Humans have a rich verbal language to express how they feel, as well as express their intentions, whereas dogs have to rely largely on body postures and facial expressions. The body language of dogs is very rich, however, and we can learn a lot about our dogs if we take the time to pay attention to the many subtle gestures and expressions they employ.

A full description of body language in dogs is beyond the scope of this book (and there is much we do not know), but we have chosen a few key examples of things to look out for so that you can identify and 'read' some important messages.

However, it is essential to remember that all body language should be interpreted in context, so take care when drawing conclusions. If you are concerned about your dog's expression in a specific situation, it's best to play safe and seek professional assistance.*

- I am worried about something

Body weight shifted backwards; hesitant to approach or actually moving away or circling around to increase distance. May approach and back off again, or bark a few short, powerful barks. Head held low and ears may move from pricked forward to pulled back, or may alternate. Tension in the muscles of the face with the eyes usually pulled into slits (he may have frown lines or a furrowed brow appearance); tail tucked under. If it's a person or another dog that is causing the consternation the tail may wag slightly, but the wagging movement is usually small, tight, and may be slow.

If your dog is worried about something that might happen rather than about what is happening now, you could see the more general posture changes listed above, but instead of a hesitant approach or avoidance, you are likely to see him scanning the environment, keeping a look out for whatever he's anxious about.

- I am very frightened

You will probably see all of the above behaviour, and in addition you are likely to see him freeze or try to run away. He may also pant and tremble, and his body will be very low to the ground,

Please note, very little scientific study has been undertaken to validate our understanding of the body language of dogs. What is presented here is the expert opinion of the authors, derived from their assessment of the available evidence, and their experience in the field of clinical animal behaviour

This smaller dog's body language indicates mild fear. Note the weight shifted backwards, tucked in hindquarters and tail, pulled back ears, and slightly narrowed eyes.

with wide open eyes that appear black as a result of pupil dilation. Frequently, his mouth will be open, but whether open or closed you will see the lips pulled back so that the mouth opening appears long, with the folds around the mouth tight and tense.

● I am not sure what to do here
Some of the most common signs include: lip licking, yawning, sniffing at the ground, scratching and 'wet-dog' shaking. Many of these signs may also occur after an arousing event, so can provide information about how a dog felt about what has just happened. However, these gestures can occur in other situations, too, so always interpret them in context.

A lip or nose lick or a yawn can all be expressions of uncertainty or conflicting desires or emotions, so do take note of when these behaviours occur (as with all body language, these must be interpreted in context). If the rest of the dog's body language confirms arousal or concern (for example, the muscle tension seen in the terrier's face leading to the frown line and lips pulled forward), gently and calmly move him away from the situation.

... in this image you can see the signs are expressed even more clearly, as he has curved his body away, and is tilting his head to avoid interaction. If the Labrador doesn't get the message soon, it would be advisable for his owner to calmly call him away.

The terrier is showing signs to tell the Labrador that he is not interested in a game. In the first image, above, note the terrier's tense face with lips pulled forward and narrowed eyes, compared with the relaxed facial expression of the Labrador ...

● I want you to go away

Dogs may want an individual to move away for a number of reasons; most commonly one of the following –

♦ he is worried about the interaction
♦ he doesn't want to lose something he has
♦ he has learnt that he can take control of a situation (not wanting to be handled, for example) by behaving in a certain way

Following are a few of the signs you may see in each of these situations. Again, they will vary with context.

▲ In the early stages when a dog is worried, the signs can be some of the most difficult to read as dogs usually start with very subtle postures and expressions. Muscles become tense and the body still, which is often paired with narrowed, blinking eyes, and possibly lips pulled back into a thin line (a form of 'grin'). A head or body turn away may frequently follow. Beyond this point you will usually begin to see some of the signs listed above under 'I am worried' and 'I am not sure what to do here' (depending on the situation). Additionally, he may lie down on his side or back (see above comment on rolling over).

If these subtle signs don't have the desired effect, they may be followed by more overt threats such as lip lifting, growling, and finally snapping, lunging, and – if he feels he has no other choice – biting to get the individual to go away. Remember, a growl is a plea for the individual to back off; not a challenge for a fight.

Although the foregoing signs have been listed as a sequence, individual dogs may use the postures and expressions in a different order, and, of course, if a dog has learnt that some gestures are ignored or get him into trouble, he may skip them totally.

The Collie is very clearly telling the Labrador not to approach his toy. Note the fixed, direct stare and the ears pricked forward. The Labrador's body language is clearly showing recognition of this, and expressing a desire to avoid confrontation (but his proximity and orientation tell us that he would be very keen to take the toy if there was any chance of doing so!).

⌃ Dogs who are defending something they want to keep (food, a possession, or their space, for example) usually start with increased muscle tension and staring, with their body weight shifted forward and their tail very tense and, frequently, raised. Again, if this is not successful, a lunge, snap or bite may follow. Likewise, if a dog has no option to escape (including if he is on a lead), he may have to proactively approach the individual with a threat to get them to withdraw. As before, this is not the dog looking for a fight, but a request that the individual move away.

⌃ Dogs who have learnt that they can successfully control a situation with one particular gesture may skip all of the signs associated with discomfort, and go straight to the gesture that has worked in the past (possibly with other individuals), which might be body weight forward, extreme muscle tension, and possibly a lowered head. He will probably stare with narrowed eyes, show lips puckered forward, possibly exposing his incisors, along with growling or snapping.

● I am frustrated
The key here is to consider the context: frustration occurs when dogs can't do something they want to because there is a barrier preventing them. This barrier may be real or perceived. Real barriers may include a gate, a door, or even a lead.

In general, dogs who are frustrated are highly excited (barking, bouncing, pacing, jumping up), and frequently show behaviours listed above under 'I don't know what to do here.'

Finally, do remember that frustrated dogs may redirect their feelings to another individual or object in the environment, and can bite innocent bystanders as a result.

● I want to greet you
Will approach with relaxed muscles, weight forward, tail usually wagging loosely and widely; facial expression relaxed without muscle tension around the eyes and mouth.

● I want to play
The play bow – front half of the body low and back half raised – is a key indicator of intent to play, and also a sign to interpret what happens next in this context. Again, natural dog behaviour includes nipping and growling to entice another to play. When directed towards people, these gestures may be inappropriate, and he may need to learn to use more suitable alternatives.

Other play gestures include a relaxed, open-mouthed play

continued page 41

Helping minds meet

These images illustrate common expressions of frustration. The lead is serving as a barrier that is preventing the dog from getting to something (either something he wants to reach or something he wants to cause to move away). He is totally focused on this object/individual, with high muscle tension in both body and face (lips pulled down over his teeth, ears pulled back and eyes narrowed), and, as is frequently seen when frustration occurs in this context, he is bouncing forward and backward at the end of the lead, and barking.

These two dogs are greeting each other very politely. They meet nose-to-nose, facial expressions showing interest as well as deference on the part of the Lurcher (pulled back ears, narrowed lips and averted gaze), and confidence without threat on the part of the Collie (direct focused gaze in the first image, which is starting to relax by the second image).

The Lurcher is trying to get the Collie to engage by offering a play bow. The Collie is disinterested and avoiding engagement, but his facial expression shows he is relaxed. So, unless the Lurcher becomes too persistent and the Collie begins to look concerned, this situation can be left to resolve itself – the Lurcher is likely to give up the solicitation quite quickly, especially if there is someone else he can play with.

Everything about the Collie who is standing is saying 'play with me.' Note the play face, body weight forward, and raised wagging tail. The behaviour of the Collie lying down cannot be properly interpreted in this still photograph as he could be crouching before engaging in a chase sequence, or the crouch and pulled back ears and lips could be an expression saying 'please leave me alone.' Owners would need to see the sequence evolve, and if the crouching Collie didn't soon join in the game, it would be better to interrupt the situation and allow the dogs to move away from one another.

Play includes a range of behaviours such as rolling over, pinning down, and biting. To interpret whether or not this is play, or one dog bullying the other, it's necessary to watch for reciprocal activity in the game (do they take turns playing 'top dog'), as well as both dogs' willingness to re-join the game if the activity is interrupted.

Not all dogs who are lying down are relaxed, although this dog clearly is. He is lying on his side, limbs outstretched, and head and tail in full contact with the ground. There is no tension in his body or facial muscles. The close-up of his face shows slightly narrowed eyes without any tension around them, and ears tilted slightly sideways.

face and a gambolling gait. However, dogs can express virtually every posture and behavioural sequence in their repertoire during a bout of play. A feature of play is that these stances and expressions are mixed into 'non-functional' sequences: ie they are not the complete behavioural sequence. For example, the dog may stalk, then run away, or chase, then roll over.

TOP TIP
To assess whether an apparently playful interaction is being enjoyed by all parties, check that roles are exchanged from time-to-time: eg all participants get a chance to chase the others. You should also note that, if the play is interrupted, all parties are keen to rejoin the game.

- I am relaxed

May be standing, sitting or lying down. Muscles visibly relaxed – mouth may be open or closed, but without tight muscle lines; ears may be swivelled sideways or drooping, and eyes not pulled wide or into narrow slits, so the muscles around them appear relaxed. The tail is in a resting position typical for the breed.

Note: Not all dogs who are lying down are relaxed – a relaxed laying position usually has the dog over on one hip, or even flat on his side. A tense position usually has him lying with his legs tucked under, and his face and muscles will show tension.

- I am interested in this

Weight forward – whether standing or sitting (often young dogs

continued page 44

All of these images show dogs expressing interest in something. The Labrador is focused, but fairly relaxed, with gaze orientated towards the object or situation of interest; ears in a semi-pricked position and tail slightly above resting position, but not raised. The first image of the terrier (right) shows an expression of interest with a degree of tension – the ears are pricked, and there is muscle tension in the face and body, with the tail raised horizontal to his back. The second image shows the same dog a few seconds later as he becomes even more focused with increased muscle tension in his body, body weight over his forequarters, and tension in his facial muscles pulling his lips forward. Ideally, this owner should direct his attention away from this situation, as the increasing arousal levels are likely to result in an outburst of unwanted emotion.

will sit to attend to something that they are trying to make up their minds about), or even lying down. Head usually up and may be tilted, ears pricked; eyes focused on object or individual of interest. No evidence of extreme muscle tension.

- I am thinking about hunting

Still, tense body posture, possibly slightly lowered but with weight forward; focused facial expression: very quiet in order not to advertise their presence to the prey. If he does then actually start to chase, the chase will be very focused and silent, constrasting with the usually looser body posture of a play-chase, where he may bark in excitement.

It is important to remember that excitement is not the same as happiness. Excited dogs usually move fast and erratically (maybe bouncing around or jumping up). They may bark or yap, and usually focus their attention on whatever it is that is exciting them. To try to determine the underlying emotion, look for the following –

- I'm excited and happy

Actions are directed at the object or individual that is eliciting the response, tail is widely and wildly wagging; mouth open in 'play face'-type expression.

- I'm excited and distressed

Tension around the eyes, mouth probably closed with tense muscles at the edges, tail tucked; actions partially focused on a third party – possibly you – although his main focus is on the situation of concern. When you see these signs, it's a good idea to help him out by calmly moving him away, and then encouraging him to refocus on something positive.

As dogs don't have the mental (cognitive) capacity to understand the differences in our world views and communication styles the way we do, it is our responsibility to adjust our behaviour and communication so that we can accommodate and assist them. In this way we will grow our ability to communicate with one another in a way which strengthens all aspects of our relationship.

This Collie is expressing happy excitement. Note the variation of the play face (all teeth showing, but lips relaxed), and raised sweeping tail as he gallops in play with the Lurcher.

WORKSHEET 4

It is useful to get into the habit of stopping and asking yourself when you see your dog moving between A and B: is there something he wants to interact with at B, or something he wants to avoid at A? Is he moving with confidence or concern? This will help you to direct his behaviour appropriately and offer him appropriate support.

Situations where my dog chooses to move from one place/person to another	*Why I think he is making this choice (based on his body language and expressions)*
. . . .	. . . .
. . . .	. . . .
. . . .	. . . .
. . . .	. . . .
. . . .	. . . .

In situations where I think he is making the choice through avoidance, is there anything I can do to make him more comfortable?

. . . .

. . . .

. . . .

. . . .

. . . .

. . . .

. . . .

. . . .

5 Communication is a two-way process

Just as we need to be able to 'read dog' and understand what our dogs are communicating to us, so, too, do we need to help them to understand us so that they can follow rules and respect boundaries, as well as comply with our requests when we ask them to do something. The previous chapter focused on the key ways in which dogs naturally communicate with others; this chapter focuses on those things we can do to ensure that we communicate with them more clearly to avoid confusion and unwanted behaviour. Dogs are extremely attentive to humans, and by watching us they learn what event predicts another and how to be well-behaved around us. While some dogs seem able to make the right choice no matter what their owner does, there are measures we can take to help us achieve the goal of a relationship without conflict.

Let's start by considering how dogs may perceive some of our natural gestures, since this can help us avoid some unnecessary problems in the relationship.

> REMEMBER!
> An obedient dog is not necessarily a well-behaved dog. An obedient dog does what we ask, whereas a well-behaved dog learns to predict our behaviour and situations, and decide to do the right thing without us needing to ask him.

Responding appropriately to your dog's signals

If we have read the messages discussed in the previous chapter correctly, we then need to respond appropriately – from the dog's perspective. Because we may want our dogs to get on with everyone and everything, it may be difficult for us to accept him avoiding someone or growling at an individual. However, if we tell him off at this point we have ignored his feelings, and not helped him to cope or change the way he feels: he still wants the individual to go away but has learnt that growling gets him into trouble. The risk in this situation is that the next time he won't growl, but may make his feelings clearer in the form of a snap or even a bite. Telling him off may also damage his confidence in our ability to keep him safe, which may make him more anxious, and/or less responsive to us. This can then become a vicious cycle, and may end up with him demonstrating a level of aggressive response for which there was

no need in the beginning, with a serious breakdown in relations between your dog and yourself as a consequence.

> REMEMBER!
> If you respond appropriately to the whisper, there's no need for him to shout.

Another common misunderstanding relates to the 'guilty' look (discussed in the previous chapter), where he is trying to defuse a potentially threatening situation in response to our signs of displeasure. The right thing to do here is to back off, because if we tell him off or punish him, we can, again, make things worse. Some dogs will become more anxious over time, and may end up using an aggressive response to defend themselves from the escalating threat.

Potential misunderstandings

Dogs can also be masters at reading the signals that we give even if we are not intending to communicate with them. Sometimes our gestures can cause them to misunderstand our intentions. We have already mentioned that certain gestures can be perceived by dogs as threatening, but there are other things we do without meaning to which can confuse or upset dogs.

Gestures which threaten

Listed below are a range of gestures that we commonly use which can be perceived as threatening by dogs, and so should be avoided, especially by anyone who does not have a strong, positive relationship with your dog.

- Looming towards or over him, and especially reaching over his head or shoulders.
- Sustained direct eye contact with him.
- Hugging and kissing.
- Playing with his muzzle with your hands.

Most dogs can learn to accept these gestures, and potentially find them enjoyable, but this is a learnt response which

This dog is turning away from the hand that is reaching over his head. This gesture, with ears and lips pulled back, is typical of a dog trying to avoid a situation which makes him feel uncomfortable or threatened.

comes from a trusting relationship and sometimes requires specific training. For this reason, it is best not to assume that even your own dog will always enjoy these actions.

- Fast and especially accelerating and direct movement towards a dog may be more threatening than steady, slow and less direct movement. This is why young children may inadvertently trigger a self-defence response, whereas an adult approaching does not.

- Erratic and unpredictable movements (from the dog's perspective), so don't be surprised if a dog who is usually good

with people reacts differently to someone who uses a walking stick or frame, or moves in an unusual manner. We accept these differences because we understand them, but to dogs they may seem threatening

Gestures which confuse

If what you are saying and what you are doing appear to contradict one another it can be very confusing for your dog – and frequently we don't even recognise that we are doing this. For example, if we are upset while trying to call our dog to us, our words may say 'come,' but the body says 'stay away.' This may confuse him and cause him to keep his distance. We consider this further below.

Gestures which encourage inappropriate responses

Dogs frequently find things that move quickly away from them exciting, and therefore they may give chase: children running around, for example. They may also show a similar reaction to quick, erratic gestures, and respond by grabbing at limbs. The natural human response, to push back, or whisk the hand or foot quickly in the opposite direction (maybe even calling out at the same time) may simply increase the excitement and potential risk in the situation.

When interacting with a young, playful dog, moving slowly and calmly around him is key. At the same time it is important to consider appropriate games where he can safely chase and grab toys to give him a safe outlet for this natural desire, and so decrease the risk of him doing so inappropriately.

Finally, we need to be careful about sighing, puffing or blowing on dogs, which can be mistaken for an invitation to play and cause a mouthy nip in response, possibly to the face.

Top Tip

Because dogs need to learn to understand us, we can make it easier for them to succeed by ensuring that our interactions with them are consistent. If we sometimes allow jumping up or grabbing at our hands or clothing in play, but get angry when this happens at other times (possibly when we are smartly dressed or not feeling well), he will become confused and/or frustrated. Confusion may give way to more general anxiety, and for some dogs this may express itself as a problem behaviour, such as an aggressive response.

Creating reliable and desirable associations

Dogs are very good at paying attention to what we are doing and predicting what this may mean for them. Just think how easily they make the decision to remain lying on their bed when you put your smart shoes and coat on as opposed to how quickly they run to the door as you don your outdoor walking gear. This understanding – and the responses dogs perform as a result of it – develops through

Learning to play tug games according to rules means that dogs can engage in an activity which most of them really enjoy while people don't risk injury. The rules of tug are: 1) wait until you are invited to take the item (in this case the dog is showing good self-control sitting while waiting for the invitation), 2) tug on the toy without grabbing at my hands, 3) let go when I ask you to, and 4) each of us wins and gets the toy sometimes. Following these guidelines keeps everyone safe, reduces the risk of the dog grabbing at just anything to tug, and finally, if he does accidentally grab something you don't want him to, you can get him to let it go without a battle of wills

their ability to learn to associate events and make predictions. Some learnt associations may relate to patterns in our behaviour that we are not even aware of, which means that they can develop appropriate and inappropriate responses equally easily: for example, the dog who starts attention-seeking barking as we end our telephone conversation because he has picked up on how our conversation tone changes at that point (and has the response reinforced as we give him attention for the action by telling him off, or trying to distract him).

A dog's power of prediction when there are reliable associations is important to understand for two reasons. Firstly, it helps us to manage our own behaviour to encourage appropriate responses from our dogs; secondly, it reminds us that if we want dogs to learn appropriate responses to our actions we need to set up reliable situations for learning. For example, if we don't want a dog who bounces around, yapping as we try to put his lead on to go out, because he has been working himself into fever pitch as we make our pre-walk preparations, then we need to ensure our behaviour teaches him that we will only go out when he is calmly sitting by the door.

The only way a dog knows how to use the information we give him to choose an appropriate behaviour is if we teach him what we want. To do this well we need to communicate –

- When we need him to do something specific (ie issue clear cues).

- What pays off and what doesn't (ie highlight the response that works by manipulating the consequences of what he does (and sometimes also the situations so that he does what we need him to), so we can reward him accordingly. Additionally ensure that incorrect responses do not result in reward).

Details of how to train dogs can be found in specific training books (see Appendix 3 for suggestions), but a few tips

> ## Remember!
> 'No' is not very helpful to dogs. Words should help guide dogs towards what they should do. The concept of 'no' is really too complicated for dogs to understand.

This dog has learnt that sitting calmly and waiting whilst food is prepared pays off. Teaching this means that you won't be bounced on when trying to deliver his meal. Realising that polite, self-controlled behaviour has a positive consequence means that he is learning a skill which can then be transferred to other situations, too.

Telling your dog 'no' when you want him to stop doing something (like seeking attention from you when you are busy) is not very useful. However, teaching him that when you tell him that 'no attention is available now' (such as by lifting your hand and giving him a verbal cue) ...

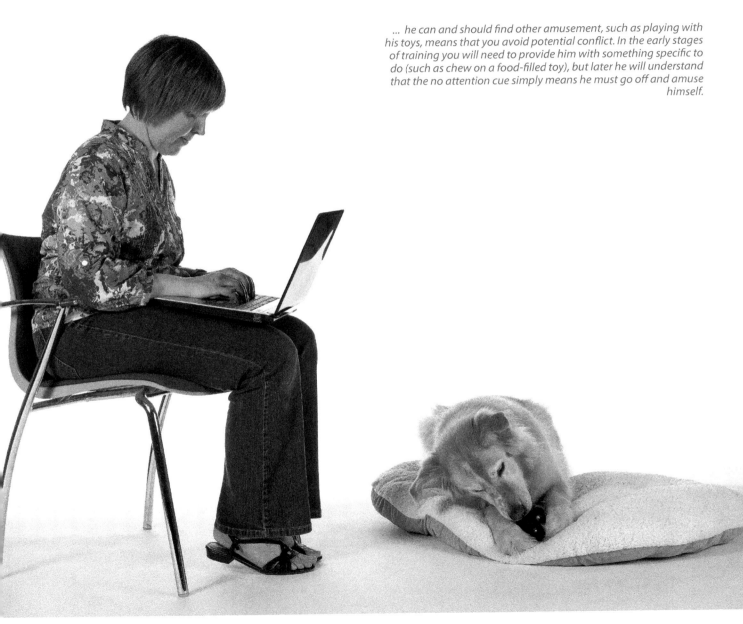

... he can and should find other amusement, such as playing with his toys, means that you avoid potential conflict. In the early stages of training you will need to provide him with something specific to do (such as chew on a food-filled toy), but later he will understand that the no attention cue simply means he must go off and amuse himself.

In these two sequences you can see that when the owner has her dog's attention she can easily ensure that he responds to her request to lie down. However ...

which can set up a situation for any instruction-based training are listed below.

Tips for getting your dog to listen to what you want

- Ensure you have your dog's attention before trying to communicate anything to him. If he is focused on something in his surroundings rather than you, he will struggle to, or be totally unable to, take in any of the information you are trying to convey.

- Regularly use your dog's name as an attention-getting device. Never use it as a reprimand. Use his name at the beginning of any request ,and wait till he has orientated to you before issuing the actual instruction.

  As soon as you get your dog, at whatever age, pair his name with something good on a number of occasions over a few days. His name then becomes a predictor of good things, and he will pay attention to you when he hears it.

  If you have adopted a dog as an adult there can be an advantage to changing his name.

- Once we have his attention we then need to encourage him to perform the behaviour that is appropriate at that time. There

... if she gives the cue when her dog's attention is elsewhere, he disengages from her and does not lie down as requested.

are many ways of doing this, but if he is not yet able to perform the action you want on your request, one of the easiest ways to get him to do so is to use an enticement (frequently a small food treat) to encourage him to take up the position or complete the action needed. For example, a small treat held at his nose and slowly moved back over his head will generally encourage him to sit.

Enticing is a good way to obtain many behaviours, as long as it is done correctly. The key point to remember is that you must fade out the enticement as soon as possible (usually after only a few repetitions) by using empty fingers to guide the behaviour, and only producing the food treat after he completes the action. In this way, he quickly learns the hand signal for the action, and you can then add a verbal cue if you wish.

TOP TIP
Don't let an enticement become a bribe. If food enticements are used for too long in the training process, you run the risk of your dog refusing

Many behaviours are easily taught using an enticement. For example, here you see a dog being taught to lie down by following a food treat held to his nose. He is rewarded with the food for moving closer and closer to the desired posture, until he can follow the treat all the way to the ground very easily. The treat is then removed so that he will follow an empty hand. This then becomes a hand signal for 'lie down.'

to comply with your request unless he first sees the food on offer. If this starts to happen, the enticement has become a bribe, and you may have a problem when no food is available, if he isn't hungry, or if something else is simply more exciting than the food you have on offer.

- What happens after an action matters. Whether we are talking about deliberately manipulated actions (such as the enticed sit), or general behaviours which your dog offers in a situation (being well-behaved), it is important to remember that what happens *after* he does something is likely to have a direct effect on whether or not he will do it again. So if, as in the scenario we talked about earlier, racing around madly is followed by going out for a walk (something he really likes), this behaviour will happen more and more. However, if every time he begins to get excited you stop your preparations, and only continue when he is calm, you are communicating to him that calm behaviour pays off.* In this sort of situation, we would suggest that you shouldn't have to ask him to sit every time; rather you should let him learn what pays off, ie look to encourage good behaviour rather than simply obedience. This makes both your lives easier in the longer term.

- Don't overdo the rewards. While it is important to reward your dog for doing the things you want him to, it is important not to overdo it. If you give rewards too often, you may end up creating a dog who finds it hard to be happy on his own. So, if your dog is quietly playing on his own with a toy, or doing some other acceptable activity, don't feel you need to tell him he's a good boy or give any additional treats. Let him learn

REMEMBER!
Feedback on what he is doing sets future expectations. If he does something you like, don't take it for granted but acknowledge it in a way that makes him and you feel good; even if this behaviour has become the norm for the situation, occasional reinforcement is important. If he does something you would rather he didn't (and doesn't need immediate intervention) make sure he doesn't get any positive feedback from you. For some dogs this includes any form of attention, including that which comes with a verbal reprimand; in this case you need to very deliberately withdraw all of your attention from him.

Please note; if patterns of behaviour are already well established, changing them can take time, and may need professional assistance. Simply withdrawing what he wants from the situation without other contingencies in place could create bigger problems, including frustration-related aggression. Please seek help from a suitably qualified behaviour practitioner if you are experiencing these problems, or if you are worried that your dog may be at risk of showing aggressive behaviour.

You don't want your dog to listen only when you have food – ie only come in from a game in the garden when you rattle the treat tin, for example. Rather, you want him to come in quickly whenever you call (you can then sometimes reward him for the behaviour afterwards). In this way you avoid teaching him that he only needs to listen to you when he sees food!

about the intrinsic pleasure of doing this sort of thing, so that he will be able to be happy and leave you alone at other times when you may be busy.

- It is really important to use words in a manner that is clear and consistent. Because dogs don't have an understanding of the meaning of words – but rather they are largely signals which predict what should happen next – we need to ensure we are very clear in our communication. For example, to us 'down' can mean 'lie down,' 'take your paws off me,' and 'get off the sofa.' Each of these actions is different, of course, so using one word for all of them will confuse your dog. It is important for the family to agree on the meaning of each word that will be used to cue your dog. In this way, he has the greatest chance of successfully responding to your requests. In other words, set up a dictionary for your dog. Using the example above, this may read –
  - ◆ Down: lie flat on the floor
  - ◆ Floor: stop jumping on me
  - ◆ Off: please get off the sofa

Helping dogs focus on verbal cues

As a verbal species, we focus on what we say to our dogs, whereas our dogs frequently focus more on our non-verbal communication. If, for example, we teach our dog to lie down by using an enticement as described above, when we fade this out he will probably still focus on our body motion more than the word we are saying. For this reason, if we need him to be able to respond to our words, irrespective of our posture – for example, if we need him to lie down as we walk past him carrying something – then we need to teach him this explicitly. This is not difficult to do: we simply need to make sure that, over a period of time, we consistently say the word BEFORE we move our body and hand so that he starts to listen to the word and anticipates the hand signal.

> ## REMEMBER!
> One word, one action.

If you don't think this is necessary, that's fine; however, you should be aware that if your dog seems to be ignoring your words, it may be because, on their own, they don't mean anything to him: he is not being disobedient; this is our problem, not his.

TOP TIP
See how well your dog understands your cues by asking for them in unusual contexts: when you are lying face down, say, or in another room (with someone else watching his reaction).

Complex requests need to be made simply

As we have already noted, words are processed very differently by dogs compared to humans. We can put words together to convey complex concepts, but this is not the case for dogs, and they will work hard to make sense of what we say, picking out from our speech familiar sounds that might contain important information. With a little forethought, we can make this easier for them.

Hints and tips for helping your dog with complex instructions

- Always ensure your instructions are clear and consistent. Your body and words must give the same message, delivered with confidence and conviction. Who comfortably follows uncertain advice? Have faith in your dog and act as though you believe he will succeed. If you are not confident he will respond, should you really be asking him to do whatever it is in this way?

- Avoid unnecessary language. Give the cues without embellishment. Saying 'please fetch the ball' is harder for a dog to understand than 'fetch,' when you've just thrown the ball.

- Deliver multiple instructions as a series of single instructions. Most dogs struggle to hold more than one piece of verbal information in their memory at a given time, so if you need your dog to follow a sequence of actions, it's best to give one instruction and then follow it with the second instruction as he gets close to completing the first action. So, if you want your dog to choose the ball from among the toys in his box and bring it to you, it is better to first say 'ball,' and then, as he picks it up, ask him to bring it to you.

- Make sure your requests don't contain contradictory information. If you change your mind about what you want, give your dog a stop signal and then ask for the new response, rather than change your instruction and expect him to understand.

> ## REMEMBER!
> If you ask your dog to do two different things at the same time, he will only do one.

Carefully considering what you and your dog are saying to each other during your daily interactions will help you to maximise the positive in your conversations, and, just as good communication improves relationships between people, you will find that this improved communication helps your relationship with your dog, too.

If your dog hasn't learnt the significance of a verbal cue, then when your hands are occupied, or in opposition to how you may have trained him, it is important that you don't get angry with him if he doesn't respond to what you ask him to do. So, in this example, if he has learnt to follow a lowered hand to lie down, it will be difficult for him to lie down when asked if you are carrying something held aloft, and he hasn't learnt to respond to the word on its own.

WORKSHEET 5

Cues my dog knows/needs to know (eg 'Down')	What this cue means to him (eg 'Lie flat on the ground')

We are all very busy, and have a tendency to make snap decisions when under pressure of time. Thinking about the consequences of our actions, or assessing all the relevant information before making a decision takes effort, and can be mentally very tiring. It also takes more time, but a little forethought can allow us to make better decisions that may prevent problems, and save time and stress in the long run. Clearly, if we analysed every potential decision we would end up paralysed and do nothing productive; we therefore need to prioritise actions and activities that deserve thought and consideration. This is true, too, for those decisions which impact on our relationship with our dogs.

Much of what we do is decided subconsciously through habit; this is the normal way of operating. The secret is to make sure that those habits which have most impact on our relationship with our dog are good ones that strengthen it. We may need to start off making very conscious decisions regarding actions which have a positive impact on our relationship, but, with repetition, what starts off as a conscious decision can soon become an effortless habit.

In this chapter we highlight some simple strategies and areas that deserve thought from us, as well as techniques for helping our dog be a little more thoughtful, since he, like you, will typically make better decisions if he takes the time to think first. Time is a precious gift that we can give each other.

> **REMEMBER!**
> The decision *not* to act is still a decision that has consequences.

Dealing with the unknown

You and your dog will inevitably encounter the unknown from time to time, whether that is a new experience for both of you – eg going on a boat together for the first time – or something that is new for one of you – eg your dog's first trip to the vet. In either case, we must be sensitive to the need for both to adapt, which means not just accepting the situation but, as far as possible, feeling confident in it.

Developing good habits, like taking regular, companionable walks in interesting new places, helps to build a positive relationship.

On a walk this dog has become a little worried by something at the water's edge. His owner allows him all the time he needs to take in information from the environment, and, after a few minutes, he can relax and continue to enjoy his walk.

An owner behaving in a relaxed manner will help her dog to relax: if you see your dog become worried about something, try hard not to tense up.

Hints and tips for managing strange situations

- Give both of you time to adjust to new situations. This means giving your dog the chance to explore in his own way as far as possible (with a loose leash and without instruction); sniffing and moving on when he feels ready to do so. Learn to read your dog's body language (see Chapter 4) to know when he is comfortable.

- Don't expect your dog to relax if you can't. As we have already noted (Chapter 5), dogs are tremendous readers of body language, and given that, as their owner, typically, you are regarded as a point of reference for safety and security, if you are nervous, your dog may respond in a similar way (he will not

REMEMBER!
Let your body language provide the reassurance that everything is okay.

If your dog is anxious in a situation, rather than trying to insist he engages in some other activity (which is frequently counterproductive), show him how much fun you are having (for example, by playing with his toys), and then, if he shows interest, make a fuss of him and continue the game with him. If he is too worried to be able to engage, change your strategy to one of the others described in this chapter.

realise that you are nervous because you are concerned about how *he* might react to this new situation). Be relaxed and, preferably, confident or even jolly (talking to yourself or indirectly to your dog about happy things is a useful exercise), but stay in control, watching for surprises.

- Let the environment provide additional reinforcement. Assuming you are relaxed, resist the urge to provide consolation to your dog, which is more often interpreted as an unusual response or sign of caution in yourself, and may make a dog anxious. Instead, it is preferable to let the environment provide the reinforcement: for example, by tossing some treats onto the ground without saying anything. He will begin to realise that progressing through the environment is associated with rewards.

- If your dog turns to you, be pleased to see him. Dogs naturally turn to their owner when they are unsure or unable to solve a problem. If your dog does this, it is an opportunity for you to provide reassurance that all is okay by being jolly about things. Tell him he's a 'good boy' in a very upbeat way, rather than try to reassure him through consolation, which is easily misunderstood by dogs as a sign of anxiety about the situation.

- If you really need to change his mind about a situation, do it by showing that you are having fun. Playing about sends a signal that all is well. This is not the same as asking your dog to play with you or pushing a toy towards him, both of which is unlikely to help an anxious dog. Rather, play a game, such as juggling some of your dog's favourite toys (with or without treats inside), and laughing as you drop them. Your dog can then choose if he wants to get involved in the game (and if he shows interest you can offer him one of the toys to play with), but it must be at his initiative rather than yours.

REMEMBER!
The more often your dog encounters strange situations and learns there is nothing to fear, the more confident he will become.

Managing impulsive behaviour effectively
Impulsivity, or acting without sufficient forethought, is a trait, and we are all impulsive to a greater or lesser degree. In some contexts it pays to be more impulsive, whilst, in others, being less impulsive may provide an advantage. This is why there is such variety in the level of impulsivity between individuals. With planning, we can put in place effective training that will help our dogs to control the

Where it is appropriate to do so – and as long as your dog is able to cope – try to take him to new places on a regular basis, as in this way he will develop his coping skills and become more resilient.

urge to act impulsively when it might be dangerous, and at the same time make good decisions, even in a hurry.

Helping your dog to respond in an emergency
Unfortunately, many owners do not think ahead to when they might need an emergency response from their dog, possibly with tragic consequences. It is always tempting to ignore the risk associated with unlikely circumstances, so we will start by considering the forethought required to make effective, fast decisions that can potentially save your dog's life.

- List all of the cues that you would like your dog to respond to immediately when you ask. For example, the 'recall' cue and 'emergency stop' can be lifesavers if you see your dog about to do something dangerous.

- When you have this list, check through and ask yourself: do I need an immediate response every time I ask for this action? If the answer is 'no', consider having the action under the control of two quite distinct words – for example, you might have one word which means 'I need you to come over here soon', and another which means 'You must come over here now!'

- Draw up a new list of actions and instructions relating to those requests that must always be complied with immediately.

- If you do not already have these (established and new) actions trained, then start training them, with the aid of a professional trainer if necessary. If your dog already has some form of response, start practising them and only reward the fastest and most complete responses (which should be the best 70 per cent or so). Set the bar according to how good your dog is at this time, raising it through the practise session. Training sessions should be short and energetic.

This dog knows how to sit when asked, so to help him learn to show self-control when he is excited, his owner has waited for him to offer to sit before she throws the toy for him.

Helping minds meet

- Although it is never necessary to shout a cue to a dog, you may have a tendency to give cues in a forceful manner if you are worried about something. It is therefore worth practising using different tones of voice so that if you do yell in an emergency you don't frighten him.

- Quit your training on a high, with a bumper reward for an excellent response.

> ### Remember!
> Better to end a training session too soon and make further progress another day, than push him too far at a given time.

Learning to control impulsivity

Being able to exercise self-control is an essential 'Life Skill'* that should be taught to all puppies, but if this hasn't happened there is no need to despair since adult dogs can also learn to be less impulsive, and by doing so make better decisions. If we give instruction to our dogs all the time, they may become very obedient, and, of course, they do need to know their boundaries (see Chapter 2), but too much instruction means a dog does not get into the habit of paying much attention to his environment, which can limit his tendency to adapt to other individuals and situations without being asked. This is likely to result in a less enjoyable relationship, and also takes away some of an individual's autonomy, which is not good for his well-being in the long term.

The key to impulse control is that appropriate and desirable behaviour is the dog's decision, rather than under the control of someone telling him what to do.

- Consider situations in which you think it is important that your dog exerts self-control – before crossing a road, when food is being prepared, when young children are playing nearby, for example.

- Draw up a list of responses that your dog knows (you can take this from your dog's dictionary – see Chapter 5), which might be appropriate in these circumstances.

- For each of the situations listed, consider what would be an acceptable response (list positive actions: eg 'to sit quietly,' and not negative statements: eg 'not to bark loudly'). Draw first from your list of known responses, and then add to this list (these additions are going to need to be trained separately). Bear in mind this is not a fixed list, and you can add to it as you think of more options.

*See Life Skills for Puppies, details in Appendix 3.

This dog is exhibiting both self-control and frustration tolerance. He has been tied up whilst his owner leaves him to lay a track for him. He is not pulling on the lead but waiting calmly (self-control), and is coping with the fact that he is restrained by the barrier of the lead from joining his owner in an activity that he enjoys (frustration tolerance). Please note: although coping with being tied up in this manner is a useful skill for dogs to learn, it needs to be carefully taught so as not to cause severe frustration, which could lead to the dog injuring himself.

- Look for opportunities to reward these behaviours in the situations described. You may need to set up some less arousing or stage-managed situations in the beginning. Ideally, so long as it is safe to do so and your dog is under control, simply wait until your dog offers one of the known responses and then reward this accordingly. In some situations it may be necessary to give a verbal prompt, but you should aim to completely fade out the cue as soon as you can.

WORKSHEET 6

The following are cues that I need my dog to respond to –

Cue .... How well he responds to it

....
....
....
....

Plans to improve responses where this is required –

....
....
....

The following are times when I would like my dog to make the decision to behave well himself, without my instruction –

Situation where I need him to make the right choice .... Behaviour I would like him to perform

....
....
....

Training I need to do to help him to make these decisions –

....
....
....
....
....

7 We both function better when we are less stressed

We all feel the effects of stress from time to time, and know that our decision-making can be impaired, and our behaviour affected at these times. For example, when we feel stressed we are often distracted, and find it difficult to focus, or even remember things, and the same applies to dogs.

So many things – a lack of focus; not listening, and being frightened by things more than usual – can all be signs that our dog is feeling the pressure. This is one of the reasons why it is so important to install and practise the emergency cues discussed in the previous chapter, in advance of any situation which might demand them.

When we are stressed we also tend to focus on the negatives in our life, and this highlights an important consideration: our mood affects the way we see things. So, if we are stressed, any misdemeanours that our dog might commit can seem so much worse, and the same holds true for him – he, too, will see his world more negatively when he is stressed.

Often, stress creeps up on us, because small stressors accumulate, even though we tend to think of stress being triggered by big events that overwhelm us. There's a plethora of self-help books offering us guidance on how to reduce the stress in their lives, but much less is written about how to evaluate the stress your dog may be under. To determine the real risk of excessive stress that your dog may be facing, you can undertake a structured assessment.

Signs of stress can be broadly divided into two groups –

- If your dog seems to be on edge and jumpy the whole time, anxious and irritable, or lick and groom himself excessively, this might indicate that there are too many things going on in his life for him to cope with: ie there are too many stressful events.

- If your dog seems depressed, has become over-attached, seems to have suddenly aged, or has frequent minor ailments – such as diarrhoea, allergies or vomiting – that do not have an obvious cause, it might indicate that your dog feels under too much pressure most of the time.

Either of the above scenarios should prompt you to consider the level of stress in your dog's life, and if you are concerned that he might be finding it hard to cope, firstly visit your vet for a chat and thorough health check of your dog.

> REMEMBER!
> Not all stress is bad, and we cannot eliminate all negative stress from our lives, but we can limit it to a level we can cope with, and learn to manage it well.

Stress auditing your lifestyle for your dog

Even if your dog is not showing any of the overt signs above, it's worth asking yourself the following questions about his daily management, routines and environment, so that you can potentially reduce the risk of excessive stress in the future –

- How would you describe your dog's personality? Is he moody? When does he get irritable?

- Is your dog under any particular form of demand at the moment (she has puppies; he is training for a competition, etc)?

- How would you describe your home? Noisy, calm, frenetic, busy, quiet, etc?

- Describe a typical day in your dog's life. Are there any dramatic shifts in your expectations from one day to the next? Are there any specific events which occur on a frequent basis that your dog finds annoying, scary or frustrating?

- Where and when does your dog tend to go to be on his own? Does he have a safe haven in your home? (See Appendix 2.)

- What sort of things does he have to put up with (ie things he has to deal with because you are unable to change them at the moment)?
  - How often do they happen?
  - What happens in these situations? How does he react?

- How does your dog cope with routine?

All healthy dogs will spend some time grooming themselves, but if you find that your dog is spending a great deal of time licking or nibbling at his fur, and your vet can find nothing specific wrong, bear in mind that this may be an indication of psychological stress.

- What sort of events happen that might affect his routine?
- How does he respond to unpredictable events?

- Who does your dog meet?
    - How would you describe the relationship between your dog and –
        - a People? (any particularly positive or negative relationships?)
        - b Other animals?

- When does your dog seek out you or anyone else?
    - How do you/they react to this?

- What rules do you have concerning the dos and don'ts of your dog's behaviour?
    - How consistently are they enforced? By each family member?

- How do you reward your dog?

- How do you let your dog know he has done something wrong?

Top Tip
Distinguish between the physical stressors (eg hard work), and psychological ones that your dog has to deal with. Try to divide the psychological stressors that you and your dog encounter into
- ▲ *Those that threaten*
- ▲ *Those that are frustrating*
- ▲ *Those that are physically painful*
- ▲ *Those that may cause sadness or loneliness*

Reducing stress for your dog
The answers to the foregoing questions should highlight the pressure points for your dog. The next stage is to identify and implement necessary changes. You might require professional support for this, but the following guidelines may help.

Everyone living with your dog needs to be involved in the process, and you should focus and agree on solutions, and not dwell on blame or guilt. You should acknowledge that some stressors may be outside your control, and so focus first on those that can be removed or reduced. However, you should later discuss how the risk of the uncontrollable stressors can be reduced. In order for things to work you need, as a family, to identify and agree on who is responsible for implementing each element of the stress management plan, and celebrate your successes and attempts at doing the right thing, even if it's not perfect each time. Be realistic in your plans and timescales (distinguishing short- from long-term

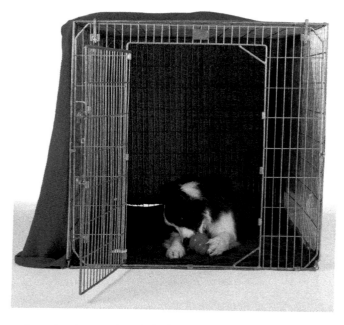

Every dog, irrespective of whether or not he may be at risk from excessive stress, should have a safe haven in his home. This is a place that has positive associations for him, where he can spend time relaxing, and where he can choose what he wants to do (see Appendix 2).

goals), not trying to do too much, too soon. It's better to do a few things well, than lots of things poorly.

Remember!
If our identification of stressors is vague, so will our plans be. The more precise and specific you can be about the issues, the clearer the solutions will be.

Hints and tips
- Make sure your dog has a safe haven in the home: somewhere he can go to where he knows he will come to no harm, and where *he* is in control. This means he will always know he can cope at home. (See Appendix 2.)

- Aim to provide a point of reference for each other in a challenging situation. For example, if he knows some behaviours that work well, ask for these as this can give your dog guidance on how he should respond in difficult situations, and become a source of positive focus for him – he knows that what he is going to do will work for him.

  Asking him to do an action he knows well and then rewarding it – for example, asking him to 'sit' and, when he does, giving a treat and/or verbal praise so he knows he has done well – is usually more successful in reducing his concerns in a situation than simply trying to distract him with a treat.

- Make sure your relationship with your dog provides a secure base. If your dog knows that no harm will come to him when he is with you, it will help him cope in even the most challenging of environments. He needs your support at all times – not your criticism.

- Supportive relationships help to build confidence. If you identify that your dog is under stress, be patient and help him. Find activities you know you both enjoy, and try to do more of them. The training can wait until another day.

- If you need to defuse a situation, do it by showing that you are having fun. As mentioned in the previous chapter, showing that you're in a mood to play sends a powerful signal that all is okay.

A word about resilience
Not only can we never avoid all stressors, but if we try to shield our dogs from everything they might perceive as negative, there is a possibility that they never have the chance to learn to cope. For this reason it is worth carefully assessing each situation where you see that your dog is feeling uncomfortable so that you can make one of the following choices –

- He is a little uncomfortable, but appears to be coping, and things are not getting worse, so I will let him figure out on his own how to manage (eg he has met a new dog on a walk and seems a little worried, but the other dog is clearly friendly and behaving appropriately).

- He is a little uncomfortable, and the situation does not seem to allow him to take positive action, so I will step in and actively support him (eg he has suddenly been surrounded by a group of unknown but friendly dogs who don't seem inclined to give him enough space, so I will distract some of them away, or step in close to him and gently guide him away).

This pup's expression says that he is not comfortable having his paw examined. His owner is offering him calm, gentle support which will assist him to feel safe and secure, and maximise the probability that he will cope with the experience.

- He is very uncomfortable, and I see no likelihood of things changing quickly, so I will remove him from the situation as quickly and calmly as possible (eg he is clearly very frightened by an extremely boisterous, unknown dog who has no intention of leaving him alone).

  In both of the latter cases, it is likely to be necessary to put

69

a specific training plan in place to help him to learn how to cope in this type of situation in the future, unless this was a one-off that he need never experience again.

Training in situations where your dog experiences stress

If your dog is under stress it is likely his learning ability will be compromised. He may be easily distracted, have difficulty identifying the associations you are trying to teach him, not be interested in the rewards you offer, or appear to forget things very quickly. However, sometimes it is necessary to train when he is under pressure, so remember that this is a time for patience and understanding on your part.

Hints and tips for training in stressful situations –

This Labrador pup is looking a little concerned about the proximity of the Spaniel, who is being carefully managed by his owner. Given time, however, he relaxes and chooses to move closer. Note the wagging tail in the second image. In controlled situations where a dog is slightly uncomfortable but it is unlikely that the situation concerning him will become more intense, allowing your dog to take things in his own time will teach him that he CAN cope, and in this way can build resilience.

- Set realistic expectations, bearing in mind each of your strengths and weaknesses. If you feel you are a long way from your goal, simply acknowledge that getting there will take time, and break this end goal into intermediate steps so that it is easier to measure progress. See this as a great opportunity for you to spend time together, growing your relationship.

- Help him to focus on the task in hand. Practising getting your dog to focus on you can be a lot of fun, as it is an opportunity for you to show your dog that you are the best thing in his world. Communicating this in increasingly distracting environments can be challenging, and may mean you need to let go of your own inhibitions to become lively enough to keep his attention on you. Don't think treats are the only, or the best, means of keeping his focus in every situation.

- Don't give him too much to do or think about at one time – break your training into tiny steps and reward every small success.

- If you know that you will also struggle to cope with the situation that your dog finds stressful, it may be best to seek professional help before you engage in training.

- Even when we are successful at coping with a stressful situation, it is demanding for all involved and we will get tired. The same is true for you and your dog. In addition, you are likely to get tired at different rates, so keep an eye on your dog and recognise his signals. If you are working on specific exercises, aim to quit before tiredness sets in, or at the earliest appropriate time after you notice the first signs of this. If you are in a less controlled situation, aim to make an exit as soon as is appropriate. This should be under your direction, so your dog learns that you are in control, that he needs to attend to

Practising having your dog focus on you when you ask him to is very useful for those times when you may really need his attention: for example, at a time when he may find things challenging. Practise focus exercises in a range of situations, and turn the act of him giving you his full attention into a fun game. Then put it 'on cue' (see chapter 5 for information on cueing) so that you can ask for his undivided attention when you need it, but also give him permission to stop focusing on you and relax and investigate his environment at other times.

you at such times, and that, if he does this, he will be safe. As you choose your exit, act decisively and with confidence so that you minimise the stress on your dog.

- With practice, resilience will build and training will become easier. Remember this when things feel challenging!

Life will always contain stressors. To minimise their impact on you, your dog, and your relationship, look for opportunities to share activities that you both love. Focusing on these positive activities – even for short periods – at times of stress will help you to support one another, and build and strengthen the positive relationship that you both want.

Make time in your day to engage in the activities you and your dog really enjoy. Watching the pleasure he derives from the time you spend having fun together will repay you for any sacrifices you need to make to give him this one-on-one time.

71

WORKSHEET 7

Situations and events which I believe my dog may find challenging (eg I have to work long hours at present so he gets fewer walks) –

1. . . . .

2. . . . .

3. . . . .

4. . . . .

Ways in which I can change these to help him (eg I can't change this for the next few months) –

1. . . . .

2. . . . .

3. . . . .

4. . . . .

Skills I can teach him to help him cope when I can't change things (eg I can provide him with a variety of interactive toys and spend time playing with him every evening) –

1. . . . .

2. . . . .

3. . . . .

4. . . . .

8 There are lots of ways we can enjoy ourselves together

Most people choosing to acquire a dog have a general idea about the relationship they expect to share. Some specifically choose a dog to participate in a canine sport, such as agility, flyball or obedience, but many simply envisage a companion who can accompany them on long walks, or snuggle up on the sofa with them in front of the television. We often underestimate the wider potential of the relationship, and the fact that we can enjoy many facets of our dog's behaviour which we may not initially have considered.

Most dogs are highly adaptable and can learn to fit into our lives very successfully, but by engaging with them in specific activities that they enjoy, we can learn more about their view of the world, as well as deepen our relationship with them. You may discover new interests and hobbies by trying some of the wide range of activities that can help to meet your dog's needs.

> **REMEMBER!**
> Whatever activities you choose to do with your dog, they should be fun for both of you. If they're not, think about an alternative way of doing the activity (possibly with a different trainer), or try an alternative activity.

There are many training classes that you can attend to learn about specific dog-related activities, and, of course, there are many books and internet sites on the subject (see Appendix 3 for some suggestions), and plenty of toys and games you can buy. However, it is perfectly possible to have fun with your dog without spending a lot of money on any of these. This section will provide you with some ideas for fun activities which cover a range of abilities (both dog and human), and includes the use of bought and home-made toys and equipment. Obviously, make sure things are safe for you both first.

Additional benefits from engaging in a variety of indoor and outdoor games with your dog include –

Most people are aware of dog agility competitions; there are also many other dog sports, so it is usually possible to find one that suits both dog and owner.

Helping minds meet

- Strengthening your dog's interest in you, so he is less likely to look for amusements which are potentially dangerous (such as chasing things) when out and about.

- Encouraging your dog to focus on you so that he is easier to direct in different and potentially challenging environments.

- The ability to keep him mentally stimulated, should you ever find yourself in a situation where going out is difficult for some reason. See also the book *No Walks? No Worries!* (details in Appendix 3) for specific information on this.

Quiet activities

Dogs who are recovering from illness or injury, as well as more elderly dogs, will particularly benefit from engaging in these sorts of games, but they are useful for all dogs. Additionally, cognitive decline in elderly dogs may be slowed if they remain mentally active, so don't neglect those older members of the family – just keep sessions short and within their ability. Puppies, too, benefit from learning that fun can be had from quiet, calm activities.

There are a range of toys, games and training activities that you and your dog can enjoy whilst remaining fairly calm and quiet.

Interactive food toys

These are toys that involve both you and your dog to varying degrees, specific examples of which include the Nina Ottosson™ range. These vary in complexity, and will encourage your dog to problem-solve in order to reach the food reward. Using his brain in this way can be very mentally tiring so keep the play sessions short at first while your dog is learning how to solve the puzzle.

Whatever toy you choose, to get most benefit from using it, your dog will need to learn how it works and be motivated to play with it. Always introduce your dog to the toy on the easiest settings and with some high value treats included inside it. Encourage your dog to investigate and work at the toy, and ensure he is getting at the food rapidly enough to prevent him becoming frustrated or abandoning the toy. As he becomes more proficient you can make this harder for him.

> **Remember!**
> If your dog is uncomfortable eating in close proximity to you, or defends his food if you approach, then do not use the interactive food toys with him. Carefully manage all interactions that involve food so as to avoid problems, and seek professional assistance with this issue.

Some food toys require interaction between dog and owner, and some are designed for the dog to play with on his own ...

... eating meals from a toy is one way of providing mental stimulation for dogs.

Games where you can vary activity levels
SCENT GAMES

You can play many different types of scent games with your dog, which may make you marvel at this wonderful ability dogs have, perceiving a world to which we are blind. Take a look at the following pictures for some inspiration.

In addition you can play games such as 'hide-and-seek' by hiding treats or food toys around the house for your dog to find. Initially, have someone gently restrain him, or ask him to sit and wait (if he knows how to) and let him watch you hide the item: behind a piece of furniture or under a cushion (if he's not likely to chew it) are good hiding places to start off with. As you release him, say 'find it' and let him go and get it.

After a few repetitions you can start to shut him out of the room before you hide the toy, and use 'find it' as you open the door and encourage him to look for it. When he gets really experienced at this, hide the toy in different rooms, or on different floors for him to search and find.

You can add scent trails based on the contents or casing of toys he will find, by dragging them to the hidden site via both direct or indirect routes.

You can also begin to add multiple treats or different items (such as a favourite toy) to the search and, if he likes bringing

... this dog has learnt to empty the toy by rocking it with his nose and paws ...

75

There are many scent games you can play with your dog. In this image the dog is searching for a treat his owner has hidden in the collection of cardboard boxes and flowerpots. This is an inexpensive means of providing your dog with mental stimulation he will really enjoy.

As well as indoor scent games there are many scenting activities you and your dog can enjoy outdoors. Most dogs enjoy tracking, and even older or less able dogs can cope with simple tracks over easy ground. Appendix 3 lists resources for teaching tracking.

things to you, you can turn it into a search, find, and retrieve game. Reward him for bringing things to you with a game played with the toy he has found (such as a game of tuggy (see page 49 for description)).

Another game most dogs enjoy is searching for a member of the family or a friend instead of a treat or toy. Reward him with a fuss or a tasty treat for finding the person who has been hiding.

> **REMEMBER!**
> To ensure that you don't over-feed your dog, keep a close eye on how much he is eating on any one day, and adjust his meals accordingly.

Tricks training
Learning new tricks is a great way to occupy your dog, and for the two of you to have fun together. You can also use these sessions to teach your dog useful skills such as to enjoy being handled for things like nail clipping.

Teaching him to go to a mat or a bed when he is asked to can be very useful, whilst teaching him to 'play dead' by lying on his side when you say 'bang' can be a fun party trick, as well as encourage calm or still behaviour. Learning to turn in a circle clockwise and anti-clockwise, or to weave through your legs are actions that many dogs enjoy, which also encourage them not to be scared by people and their gestures.

The possibilities are limited only by your imagination and your dog's ability at the time you introduce a new trick (physical and mental capacity for learning complex actions is likely to build with experience), and desire to play this new game.

There are several good training books which will give you step-by-step instructions on how to teach various tricks, and Appendix 3 has a list of suggestions for you.

Principles of trick training
- In the beginning train in short sessions with few distractions.

- Start with easy actions and build complexity over time.

- You may wish to use a food enticement initially to encourage your dog to perform the action (but do remove this quite soon so that he doesn't learn to depend on the presence of food to perform the action).

> **REMEMBER!**
> It should be fun for both of you. If you find yourself becoming frustrated, or if your dog is not engaging, leave it for the day and work out what you can change to make it easier and more fun next time, or try a different trick: there is always something else to do.

HELPING MINDS MEET

The following photo sequence demonstrates a dog learning to perform two simple tricks: 1) resting his head on his front leg, and 2) placing his chin flat on the ground between his front paws. Both tricks begin with him lying down, then rewarding him for dropping his head slightly at first, and then further and further towards the ground. Where the rewards are placed as his head drops will influence the end action, so, in the first sequence (images 3 & 4), he is fed rewards over his leg, which encourages him to drop his head onto his leg, while in the second sequence (images 5, 6 & 7), he is fed his rewards in the middle between his front legs to encourage him to drop his head between them. In both cases the rewards are fed low to the ground to encourage a lower head position. Each action can then be named so that he can be asked to perform them on cue. Training books, such as those mentioned in Appendix 3, give guidance for training a range of tricks.

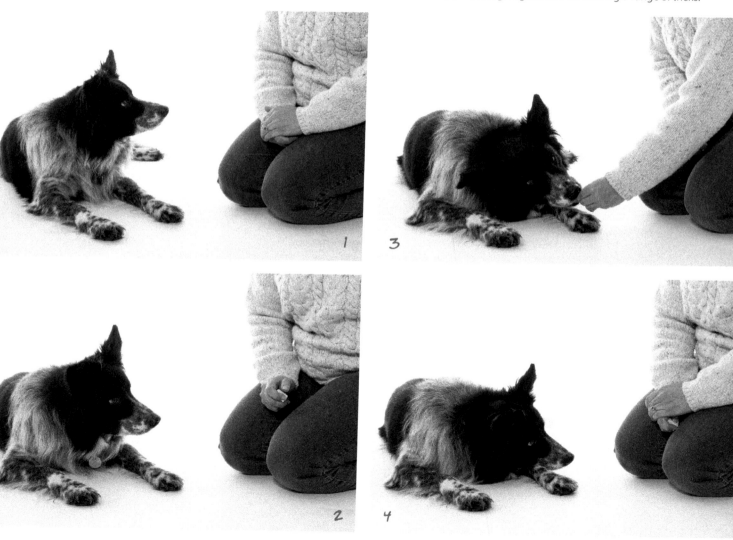

5

7

6

- The cue word (eg 'nose') or hand signal is added right at the end of your training, once the behaviour is perfect.

TOP TIP

A cue, or command, is a verbal or visual signal that prompts your dog to perform an action or behaviour: eg 'sit' means 'place your bottom on the floor.' The word or signal you choose has no meaning to your dog until he has associated it with a particular behaviour. Often, you will use a visual signal to prompt the behaviour at first: eg a hand moved in an arc to encourage your dog to turn in a circle. To add a verbal cue, teach the behaviour you want without a verbal cue at first, and only introduce the verbal cue when the behaviour is perfect.

Energetic fun

It is possible to engage in energetic activities without signing up to specific training classes. However, it is really important that safety is always considered before starting this type of work with your dog: his safety, your safety, and the safety of all those around you.

- Reward your dog immediately he performs the action you're teaching.

- Break the behaviour into small chunks (specific acts) if necessary that can be joined together later.

REMEMBER!

Interactive activities that involve all members of the family as well as friends help your dog to develop positive relationships with a range of people.

For example –
- Is the surface underfoot sufficiently secure? Slipping can cause serious joint and back injuries for both people and dogs.

- Is the area perimeter sufficiently secure? Your dog may get excited and forget to pause at the roadside.

- Is it safe to engage in your proposed activity if other people or dogs are around? Other dogs may not like your dog running past them too quickly, and it is possible that thrown toys can spark conflict between dogs. Is your dog sociable and calm with other dogs, even if they get excited by what is going on?

If you want to explore some of these activities but are worried about the security of the environment, you may be able to find a local dog training or behaviour facility that has a secure paddock or hall that you can hire on an occasional or regular basis – it's certainly worth asking.

Here are some ideas for active games which can be played in your garden, out on walks, or in a secure open space.

HOME AGILITY/OBSTACLE COURSES

You can buy a range of inexpensive agility equipment, or you can make it yourself – brooms and mops balanced on top of bricks, buckets or cardboard boxes can make obstacles for your dog to jump over; large cardboard boxes (placed open on the floor) can be ideal for him to sit inside until asked to move. Boxes with their bottoms opened up and placed on their side can form tunnels to crawl through.

A row of plastic bottles with sand in to weight them can form a line of objects to weave in and out of. You are limited only by your imagination and safety considerations, so watch some agility videos on the internet, and design your own fun course in your garden or hallway.

IMPORTANT PRACTICAL CONSIDERATIONS
- Never make the obstacles so secure that they can't be knocked over by your dog– if a stick can be knocked off its support, it is less likely that a dog will trip or injure himself.

- To minimise injury risk make sure any stick ends that your dog could run on to are protected by something, such as an old ball with a slit in it.

- Always make obstacles very easy to begin with – for example,

You can buy inexpensive agility equipment to set up small agility courses in your garden.

Alternatively, a broom or mop balanced on bricks or books can create small jumps. Always ensure that they are safe to prevent injury.

Above: A sturdy plank balanced on bricks or books can serve as a mini dog walk. By first using a food lure you can teach him to carefully balance as he walks over it. Exercises such as this can build his physical fitness and suppleness as well as his confidence. Do be sure that he doesn't focus on the treats in your hand so much that he doesn't look where he is putting his feet!

Even cardboard boxes can form part of a home-made agility course – he can be taught that when he gets to the box he should climb into it and sit inside.

make the jumps lower than your dog can easily cope with. This way he will gain confidence and build muscle strength slowly, reducing the risk of injury.

- Encourage him to negotiate the obstacles carefully, rather than simply as fast as possible. The fun and stimulation lies in the skills developed, and if he learns care and skill then the sequences can gradually be made more difficult, and he can be challenged more appropriately in this way, whilst, of course, enjoying the buzz of speed as it becomes safer for him to go faster.

RETRIEVE GAMES
Most dogs enjoy a good game of fetch. Retrieve games can be played indoors (if the flooring is non-slip and you can safely roll or toss toys), in the garden or out on walks. To build additional mental skills into these games, try the following –

- Add a search component – toss the toy into long grass or behind some bushes so that he has to use his nose to search for it.

- Add a 'wait' instruction before he can fetch the toy – this can even be extended to a range of actions he needs to perform (sit/down/give paw) before he runs off to collect the toy. In addition to making the game more fun, this helps him learn to focus on and respond to you, even when he is excited. It also teaches him the essential Life Skill "I can't have everything I want when I want it" (*Life Skills for Puppies*).

- If you can't throw very far, increase the distance he has to run by having him sit next to you and waiting (practising self-control) while a family member or friend walks a distance away from you, and throws the toy for him to fetch and bring to you. You can increase the difficulty by having him wait after the toy is thrown before you tell him to fetch it. He then has to remember where the toy went, and, if the grass is even slightly long, use his nose to find it.

Active training games
Most training can take place outdoors, but outdoor games that are especially fun include –

- Calling him between two family members, or you and a friend, or asking him to wait and then walk away from him and call him. Reward him with a game or treat when he gets to you – this strengthens his recall cue.

This sequence of images demonstrates how a simple retrieve game can be built on to allow your dog to practise different skills, and gain more from the activity. In this game he is asked to wait while the toy is thrown into longer grass, and, once the toy has fallen, is then sent to fetch it. Because of the length of the grass he has to use his nose to locate the toy, which he then returns to his owner for a game of tug. This sort of game will tire him out mentally as well as physically, and help him practise key skills such as listening to his owner, even when he is excited.

HELPING MINDS MEET

This large log is a perfect natural obstacle for a dog to enjoy a confidence and physical co-ordination building activity. The older Collie is already very confident and will happily lie on, walk along ...

- Active scent games, including more structured games like tracking (see Appendix 3 for suggested books).

- Natural agility – incorporate natural obstacles into your walk. For example, if you are in a wooded area, encourage him to jump over or walk along fallen tree trunks. Or ask him to jump onto a stump or broad tree trunk. and perform some of his tricks whilst balancing there. Learning to negotiate different surfaces underfoot and weave through vegetation helps build your dog's confidence and body awareness, and, if done carefully, his muscle strength and flexibility, too. Never encourage flat-out running through areas with rabbit holes or similar as he can slip and break a leg as a result. Controlled fun is generally better for him, psychologically, too.

TOP TIP
If there are aspects of the environment that impact on your dog's enjoyment of walking and exploring (for example, a severe fear of noises if you live in an area with bird-scarers), it is worth seeking help

... and jump over the log.

The young Labrador, however, is less confident (you can see how she has splayed her toes to grip onto the wood), and is only prepared to sit on the log and go part way into the 'down' – taking up a 'bow' position to her owner's treat. With practice, she, too, will develop the confidence and coordination to perform a number of actions on the log. Most dogs thoroughly enjoy this type of outdoor activity, and it's great fun for owners, too!

from a suitably-qualified behaviour practitioner. See the potential to overcome this as a way of building a stronger partnership

Finally, don't forget that even a walk where none of these activities are included, but he has a bit of time to explore on his own, is something most dogs really enjoy. The physical activity is only part of the fun; a large portion of the enjoyment of the walk is in exploring the environment. For this reason it is essential to train your dog to be safe off the lead so that he can fully enjoy this freedom to explore.

However, whether he is on-lead or off, allowing him to fully engage with aspects of the environment that he wants to look at, listen to or sniff will benefit his mental and emotional well-being. Watching him do this; noticing the things that attract his attention, can add to your pleasure in the environment, teach you more about your dog, and deepen your relationship at the same time.

WORKSHEET 8

Indoor activities that I would like to try out with my dog (eg hide and seek games) –

. . . .
. . . .
. . . .
. . . .

Things I need to put in place to make this possible (eg I need to find out which toys he likes best so that he is motivated to look for them) –

. . . .
. . . .
. . . .
. . . .

Outdoor activities that I would like to try (eg making his retrieve more challenging to teach him additional skills) –

. . . .
. . . .
. . . .
. . . .

Things I need to put in place to make this possible (eg I need to read up about how to teach him to stay while I throw the toy) –

. . . .
. . . .
. . . .
. . . .

Appendix 1 Loose lead walking with 'sniff' cue

It is much more pleasant to walk your dog on-lead if he doesn't pull, and walking on a loose lead is more comfortable for him, too. However, it can take a while for him to master the skill of walking at our pace and dividing his attention between the world he finds fascinating – especially its scents – and the human at the other end of the lead.

There are many different ways to teach dogs to walk nicely on a lead, and, as long as the technique you choose is not detrimental to your dog's welfare, it's up to you to determine the method that best suits you both. Regardless of the techniques you've used before, if you are struggling, it's worth starting afresh – try swapping the side that your dog walks on initially to make it new for him – and be sure to be consistent in your approach.

During your period of training, there will be times when you and your dog just need to get from A to B, and you don't have the time or the patience to be consistent with loose lead walking, so a different harness or collar is a good way of differentiating this for your dog. Choose the collar or harness you want to use when you are training your dog on a loose lead, and use a different type for times when the rules will not be in play.

The following is a suggestion for a training technique which we find works well. The aim is to enable your dog to walk with you without pulling, whilst still enjoying the walk and the environment around him. Of course, you should *both* enjoy the walk, as we regard loose lead walking as a partnership between person and dog: a contract in which your dog's undertaking is not to pull and not to trip you up, and yours is to allow your dog to enjoy the walk by doing things he enjoys – for example, sniffing bushes, investigating holes, and greeting people and dogs where appropriate. This technique is not intended to teach obedience heelwork.

- Begin with an appropriately fitting flat collar, harness or head halter attached to a lead which is long enough to allow your dog to walk comfortably alongside you with slack in the lead between your hand and him.

- Decide on which side you would like your dog to walk.

- Arm yourself with tasty, easily swallowed treats – cheese, chicken, hot dog – that are cut up very small, and adjust his daily ration accordingly.

- Find a quiet, distraction-free, safe environment in which to work. It is ideal to start in your living room or garden at first, and not after he has just been fed.

- Attach the lead to your dog's collar or harness and stand still.

- Say nothing, and as soon as he pays you any attention, tell him he's good and feed him a treat.

- Now use a treat in your fingers to bring him into position at your side.

- Make sure the lead is loose and that there is no pressure on his collar/harness: the lead is simply a safety line, not a means of directing his movement.

- Feed him treats simply for standing next to you. He is learning that being at your side is a good place to be.

- Take one step forward and, as he moves with you, feed a treat into his mouth. Try to feed him for staying alongside you; don't wait for him to get ahead. If he is reluctant to move as you step forward, use another treat to entice him into position next to you again.

- If he moves along keeping pace with you, feed him a couple more treats as you slowly take a few steps forward.

- After practising this for a few days, walk a little further before feeding a treat, so that it isn't quite as easy for him.

- If, at any time, he gets ahead of you or becomes distracted and pulls sideways or backwards, stand still immediately, don't say anything, and wait for his attention to return to you. Once you have this, use a treat in your fingers, or a hand gesture, to return him to your side, and have him walk a couple of steps next to you before receiving a treat.

Remember, he doesn't have to be glued to your leg, just moving with you and not pulling.

Take care not to feed him a treat immediately after he comes back to you as this can very quickly teach him that the quickest way to get a treat is to pull away and then return.

- Repeat this initial training in different places, with gradually increasing distractions. Practise until he can walk alongside you for about 20 paces without pulling before receiving a treat, and if he does pull, can return to position to be able to move forward without requiring a hand signal to prompt him: ie he can work out what he needs to do to move forward again.

- At this point, for most dogs, moving forward becomes a more important reward than the food, so you can try to fade out the food rewards, unless in a very distracting place where he will need extra help to get it right.

- When you reach this stage of training, you may find that he develops the very annoying habit of yo-yo-ing on the lead; pulling and then coming back so that your walk becomes a stop/start affair. In this case, the next part of the training becomes very important.

- Next time your dog pulls and then moves back into position, DON'T continue forward. You'll probably find that he anticipates you're going to move and bounces forward to pull again. Stand still.

- At this point you want your dog to understand he has to be with you mentally before he can move forward. In other words, he has to be paying you real attention, and be in tune with you, to be able to move forward, not focusing totally on the environment and just flicking you an occasional look.

- Judging this mental state can be difficult, but look for signs that your dog has moved his attention from the environment and firmly back to you. Some dogs demonstrate this by sitting next to you, some show a more relaxed posture instead of being intently focused on the wider environment, some make sustained eye contact with you.

- When you see this change in behaviour, tell him he is very good and move forward again.

- It may be beneficial to use food rewards again for a while so that the change in mental attitude receives two payoffs – food and movement – but DON'T give in to the temptation to use

a food treat to move him into position. At this point we want your dog to take responsibility for walking without pulling, which means that it will no longer be necessary for you to manage him or remind him as he will have learned to manage himself.

- With time he will develop the skill of paying you just enough attention to match his pace to yours and change direction with you while enjoying his surroundings. At this point he won't be actively looking at you at all (or rarely), but will simply be keeping you in his peripheral vision.

- To easily switch him between being allowed to sniff at his own pace and move along with you without sniffing when you need him to, you need to add additional steps to your training (see images). Once he has learnt the basics, give him a specific 'let's go' cue word as you start out on your walk. Keep his focus on you by talking to him, and possibly feeding him a few treats.

This dog is walking nicely at his owner's side, and she is engaging with him to encourage him to walk next to her.

He is given permission to investigate something of interest with both a verbal "go sniff" cue and a hand signal.

His owner walks along slowly with him at his pace so that he can investigate the interesting smells as much as he likes.

- When you approach an area that you know he will like to sniff, give him a specific 'sniff' cue and with your hand indicate towards the ground.

- Allow him to sniff for a while.

- After a time – either pottering on at his pace or standing in one spot while he has a good sniff – give him his 'lets'

go' cue again, entice him back to you and keep his focus once again.

- With time, he will learn to distinguish the two different walking instructions, and you will find that you can more easily switch him between increased focus on you and more focus on the environment.

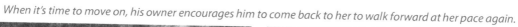

When it's time to move on, his owner encourages him to come back to her to walk forward at her pace again.

APPENDIX 2 Creating a safe haven

A safe haven is a place your dog can go to where he can feel safe and secure: a place where he is always in control, and will know that no-one will enforce anything on him whilst he is there.

To create a safe haven, find a space in your home where you can place a permanent bed for your dog. The bed can be a crate, a basket-type bed, a cushion, dog-duvet ... whatever type of bed he likes best.

Ideally, situate the bed in an area that is fairly quiet and peaceful for him, but not isolated from the family, where he is comfortable spending time, and that will be available to him at all times when he is at home. An Adaptil™ diffuser plugged in nearby may also help to create a sense of security.

Once you have positioned the bed, you can make it appealing to him by occasionally placing favourite toys, chews and treats there, when he is away from the area. In the beginning, do this frequently, but once he is using the bed regularly it is worth still occasionally placing special treats there (an especially loved or new toy, or a food treat or chew). This is one way of providing him with fun surprises which will encourage a positive emotional state, and help to counteract the stresses of daily life. We all like and benefit from good surprises in our lives!

Remember!

- Never send him to his safe haven as a 'time out' or punishment – this area must always be a place that makes him feel good

- Never make him leave his safe haven (unless it's an emergency, and even then try to ensure that you call him out of it rather than force him out)

- Never approach him in his safe haven – this must be a place he can go to and feel secure if he wants to avoid contact with people. If you are thinking of taking him for a walk, when he is in this area, ask him and respect his response. If he stays put; don't take him out. This is his way of saying 'thanks, but no thanks'

- If another dog approaches him when he is in his safe haven, quietly call them away so that, again, your dog knows he can avoid interacting with them when he is there

- Only those able to understand the rules of the safe haven should be left unsupervised around your dog and the safe haven area. Young children must never be left unsupervised around dogs

- Note when he uses his safe haven, and if he seems to be choosing it to be alone, reflect on the reason why; respect and learn from this

Appendix 3 Resources

Books

Life Skills For Puppies: Laying the foundation for a loving, lasting relationship • Helen Zulch & Daniel Mills • Hubble & Hattie

No Walks? No Worries: Maintaining wellbeing in dogs on restricted exercise • Sian Ryan & Helen Zulch • Hubble & Hattie

How Dogs Learn • Mary Burch & Jon Bailey • Howell Books

10-Minute Dog Training Games • Kyra Sundance • Quarry Books

Smellorama! Nose games for dogs • Viviane Theby • Hubble & Hattie

The Canine Kingdom of Scent: Fun Activities Using Your Dog's Natural Instincts • Anne Lill Kvam • Dogwise

Tracking From the Ground Up • Sandy Ganz & Susan Boyd • Show-me Publications

Websites
www.dogstardaily.com

Rogues' Gallery: The Cast

Spook

Tyke

Honey

Juno

Lexi

Pan

Toffee

Brimo

Frankie

Riley

Snowy

uMoya

Cooper

Ivo

Index

MORE GREAT BOOKS IN THIS SERIES!

Puppy education from the puppy's perspective! Presenting the key skills that a dog needs to cope with life, this groundbreaking book, written by professionals in the field, aims to assist owners develop a fulfilling relationship with their puppy, helping him to behave in an appropriate manner and develop resilience, whilst maintaining good welfare.
The skills taught are incorporated into everyday life so that training time is reduced, and practising good manners and appropriate behaviour become a way of life.

205x205mm • 96 pages • 121 colour images • paperback plus flaps • ISBN 9781845844462 • £12.99*

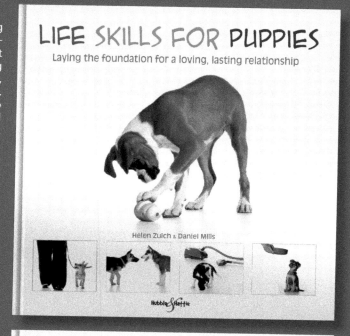

Meeting a dog's physical, mental, and emotional needs during a period of limited mobility can help reduce the possibility of future behaviour problems, alleviate some of the stress of caring for a less active canine, and help aid recovery. Encouraging owners to reflect upon, and take into account, their dog's individual requirements in advance of surgery or other lifestyle-limiting event, the book also contains information and advice about appropriate activities that owners can introduce to their dog's daily routine whilst walks are limited

205x205mm • 96 pages • 100 colour images • paperback plus flaps • ISBN 9781845846053 • £12.99*

For more info on Hubble and Hattie books please visit www.hubbleandhattie.com; email info@hubbleandhattie.com; tel 44 (0) 1305 260068.
*prices subject to change